[1]

Python Unlocked

Become more fluent in Python—learn strategies
and techniques for smart and high-performance
Python programming

About the Author

Arun Tigeraniya has a BE in electronics and communication. After his
graduation, he worked at various companies as a Python developer. His main
professional interests are AI and Big Data. He enjoys writing efficient and testable
code, and interesting technical articles. He has worked with open source technology
since 2008. He currently works at Jaarvis Labs Limited, India.

I would like to thank my parents and elder siblings, Ashok and
Asha, who have always supported me in completing this book with
good quality. A special thanks to the people at Packt for being so
understanding even though I missed a few deadlines.

About the Reviewers

Mike Driscoll has been programming in Python since 2006. He enjoys writing about Python on his blog at http://www.blog.pythonlibrary.org/. He co-authored *Core Python Refcard for DZone*. Mike has also been a technical reviewer for *Python 3 Object Oriented Programming*, *Python 2.6 Graphics Cookbook* and *Tkinter GUI Application Development Hotshot*, all by Packt Publishing. He recently wrote the book *Python 101*, and is working on his next book.

> I would like to thank my beautiful wife, Evangeline, for always supporting me. I would also like to thank my friends and family for all that they do to help me. Finally, I would like to thank Jesus Christ for saving me.

Chetan Giridhar is a Python developer, open source enthusiast, and start-up specialist. He has authored/reviewed books on Python, published papers in journals, and delivered talks in conferences around the world. You can get in touch with him at cjgiridhar@gmail.com.

He has also reviewed *IPython Interactive Computing and Visualization Cookbook* by Packt Publishing.

> I would like thank my mentors, friends, colleagues, and my ever-supporting family.

Sanjeev Jaiswal is a computer graduate having 6 years of industrial experience. He uses Perl, Python, and GNU/Linux for his day-to-day work. Sanjeev teaches Perl, Python, web development online as well. Sanjeev has worked closely with major clients, such as CSC, IBM, United Online, and Syniverse. He has also developed an interest in web application security since 2013.

Sanjeev loves teaching technical stuff to engineering students and IT professionals, and he has been teaching since 2008. He founded Alien Coders (http://www.aliencoders.org/) based on the *Learning through sharing* principle for computer science students and IT professionals in 2010, and it became a huge hit in India amongst engineering students.

He usually uploads technical videos on YouTube under the *AlienCoders* tag. One can follow him on:

- Facebook at http://www.facebook.com/jassics
- Twitter at http://twitter.com/jassics

He wrote an Instant book called *WebSpeed Optimization Techniques*, and *Learning Django Web Development* by Packt Publishing, and he is always looking forward to author or review more and more books from Packt and other publishers.

I would like to thank my parents and my wife, Shalini Jaiswal, for moral support at every phase of life and growth. I also give deep thanks and gratitude to my best friends Ritesh Kamal and Ranjan Pandey for their personal and professional help all the time.

It is because of them and other friends that I learned and achieved a set of impossible goals in a short time.

Vishrut Mehta is a dual degree student of IIIT Hyderabad, currently in his 5th year. He is completing his MS under Dr. Vasudeva Varma, Research Dean of IIIT, at Search and Information Extraction Lab (SIEL). He works in areas related to software-defined networks and cloud computing. He completed his internship at Google last summer and a research internship at INRIA, France, in 2014. During his internship, he worked on various areas of cloud computing, such as automatic reconfiguration of a multicloud application. Vishrut also participated in Google Summer of Code in 2013 and was also was the admin for the Google Code-in between 2013 and 2014 for Sahana Software Foundation.

He has also reviewed *Learning Python Network Programming, Untangle Network Security, and Python Network Programming Cookbook*, all by Packt Publishing.

I would like to thank my guides Dr. Vasudeva Varma and Dr. Reddy Raja for constantly supporting my ideas and helping me in my work. I would also like to thank my parents who never lost their faith in me.

www.PacktPub.com

Support files, eBooks, discount offers, and more

For support files and downloads related to your book, please visit www.PacktPub.com.

Did you know that Packt offers eBook versions of every book published, with PDF and ePub files available? You can upgrade to the eBook version at www.PacktPub.com and as a print book customer, you are entitled to a discount on the eBook copy. Get in touch with us at service@packtpub.com for more details.

At www.PacktPub.com, you can also read a collection of free technical articles, sign up for a range of free newsletters and receive exclusive discounts and offers on Packt books and eBooks.

https://www2.packtpub.com/books/subscription/packtlib

Do you need instant solutions to your IT questions? PacktLib is Packt's online digital book library. Here, you can search, access, and read Packt's entire library of books.

Why subscribe?

- Fully searchable across every book published by Packt
- Copy and paste, print, and bookmark content
- On demand and accessible via a web browser

Free access for Packt account holders

If you have an account with Packt at www.PacktPub.com, you can use this to access PacktLib today and view 9 entirely free books. Simply use your login credentials for immediate access.

Table of Contents

Preface

Python is a versatile programming language that can be used for a wide range of technical tasks—computation, statistics, data analysis, game development, and more. Though Python is easy to learn, its range of features means there are many aspects of it that even experienced Python developers don't know about. Even if you're confident with the basics, its logic and syntax, by digging deeper you can work much more effectively with Python—and get more from the language.

Python Unlocked walks you through the most effective techniques and best practices for high performance Python programming—showing you how to make the most of the Python language. You'll get to know objects and functions inside and out, and will learn how to use them to your advantage in your programming projects. You will also find out how to work with a range of design patterns, including abstract factory, singleton, and the strategy pattern, all of which will help make programming with Python much more efficient. As the process of writing a program is never complete without testing it, you will learn to test threaded applications and run parallel tests.

If you want the edge when it comes to Python, use this book to unlock the secrets of smarter Python programming.

What this book covers

Chapter 1, Objects in Depth, discusses object properties, attributes, creation and how calling objects work.

Chapter 2, Namespaces and Classes, discusses namespaces, how imports work, class multiple inheritance, MRO, Abstract classes, and protocols.

Chapter 3, Functions and Utilities, teaches function definitions, decorators, and some utilities.

Chapter 4, Data Structures and Algorithms, discusses in-built, library, third party data structures and algorithms.

Chapter 5, Elegance with Design Patterns, covers many important design patterns.

Chapter 6, Test-Driven Development, discusses mock objects, parameterization, creating custom test runners, testing threaded applications, and running testcases in parallel.

Chapter 7, Optimization Techniques, covers optimization techniques, profiling, using fast libraries, and compiling C modules.

Chapter 8, Scaling Python, covers multithreading, multiprocessing, asynchronization, and scaling horizontally.

What you need for this book

You should have a working installation of Python, preferably greater than 3.4. You can use this with Python 2 as well but the book uses Python 3 and introduces its many new features.

Who this book is for

If you are a Python developer and you think that you do not fully understand the language, then this book is for you. This book will unlock mysteries and reintroduce the hidden features of Python to write efficient programs, making optimal use of the language.

Conventions

In this book, you will find a number of text styles that distinguish between different kinds of information. Here are some examples of these styles and an explanation of their meaning.

Code words in text, database table names, folder names, filenames, file extensions, pathnames, dummy URLs, user input, and Twitter handles are shown as follows: "So, we can change an object's type by changing its `__class__` attribute."

A block of code is set as follows:

```
def __init__(self, name):
        self.name = name
        self._observers = weakref.WeakSet()

    def register_observer(self, observer):
        """attach the observing object for this subject
        """
        self._observers.add(observer)
        print("observer {0} now listening on
            {1}".format( observer.name, self.name))
```

When we wish to draw your attention to a particular part of a code block, the relevant lines or items are set in bold:

```
self.assertFalse(assign_if_free(mworker, {}))

    def test_worker_free(self,):
        mworker = create_autospec(IWorker)
        mworker.configure_mock(**{'is_busy.return_value':False})
        self.assertTrue(assign_if_free(mworker, {}))
```

New terms and **important words** are shown in bold. Words that you see on the screen, for example, in menus or dialog boxes, appear in the text like this: "Let's take an example of an object **iC** instance of the C class with the **str** and **lst** attributes."

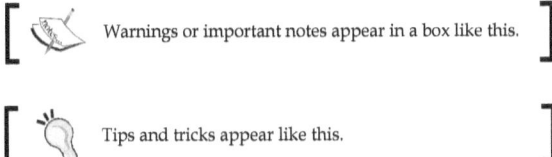

> Warnings or important notes appear in a box like this.

> Tips and tricks appear like this.

Reader feedback

Feedback from our readers is always welcome. Let us know what you think about this book—what you liked or disliked. Reader feedback is important for us as it helps us develop titles that you will really get the most out of.

To send us general feedback, simply e-mail feedback@packtpub.com, and mention the book's title in the subject of your message.

If there is a topic that you have expertise in and you are interested in either writing or contributing to a book, see our author guide at www.packtpub.com/authors.

Customer support

Now that you are the proud owner of a Packt book, we have a number of things to help you to get the most from your purchase.

Downloading the example code

You can download the example code files from your account at http://www.packtpub.com for all the Packt Publishing books you have purchased. If you purchased this book elsewhere, you can visit http://www.packtpub.com/support and register to have the files e-mailed directly to you.

Errata

Although we have taken every care to ensure the accuracy of our content, mistakes do happen. If you find a mistake in one of our books — maybe a mistake in the text or the code — we would be grateful if you could report this to us. By doing so, you can save other readers from frustration and help us improve subsequent versions of this book. If you find any errata, please report them by visiting http://www.packtpub.com/submit-errata, selecting your book, clicking on the **Errata Submission Form** link, and entering the details of your errata. Once your errata are verified, your submission will be accepted and the errata will be uploaded to our website or added to any list of existing errata under the Errata section of that title.

To view the previously submitted errata, go to https://www.packtpub.com/books/content/support and enter the name of the book in the search field. The required information will appear under the **Errata** section.

Piracy

Piracy of copyrighted material on the Internet is an ongoing problem across all media. At Packt, we take the protection of our copyright and licenses very seriously. If you come across any illegal copies of our works in any form on the Internet, please provide us with the location address or website name immediately so that we can pursue a remedy.

Please contact us at copyright@packtpub.com with a link to the suspected pirated material.

We appreciate your help in protecting our authors and our ability to bring you valuable content.

Questions

If you have a problem with any aspect of this book, you can contact us at questions@packtpub.com, and we will do our best to address the problem.

Objects in Depth

1

In this chapter, we will dive into Python objects. Objects are the building blocks of the language. They may represent or abstract a real entity. We will be more interested in factors affecting such behavior. This will help us understand and appreciate the language in a better way. We will cover the following topics:

- Object characteristics
- Calling objects
- How objects are created
- Playing with attributes

Understanding objects

Key 1: Objects are language's abstraction for data. Identity, value, and type are characteristic of them.

All data and items that we work on in a program are objects, such as numbers, strings, classes, instances, and modules. They possess some qualities that are similar to real things as all of them are uniquely identifiable just like humans are identifiable by their DNA. They have a type that defines what kind of object it is, and the properties that it supports, just like humans of type cobbler support repairing shoes, and blacksmiths support making metal items. They possess some value, such as strength, money, knowledge, and beauty do for humans.

Name is just a means to identify an object in a namespace similar to how it is used to identify a person in a group.

Identity

In Python, every object has a unique identity. We can get this identity by passing an object to built-in ID function ID (object). This returns the memory address of the object in CPython.

Interpreter can reuse some objects so that the total number of objects remains low. For example, integers and strings can be reused in the following manner:

```
>>> i = "asdf"
>>> j = "asdf"
>>> id(i) == id(j)
True
>>> i = 1000000000000000000000000000000000
>>> j = 1000000000000000000000000000000000
>>> id(j) == id(i) #cpython 3.5 reuses integers till
256 False
>>> i = 4
>>> j = 4
>>> id(i) == id(j)
True
>>> class Kls:
...pass
...
>>> k = Kls()
>>> j = Kls()
>>> id(k) == id(j) #always different as id gives memory
address False
```

This is also a reason that addition of two strings is a third new string, and, hence, it is best to use the StringIO module to work with a buffer, or use the join attribute of strings:

```
>>> # bad
... print('a' + ' ' + 'simple' + ' ' + 'sentence' + ' ' + '')
a simple sentence
>>> #good
... print(' '.join(['a','simple','sentence','.']))
a simple sentence .
```

Value

Key 2: Immutability is the inability to change an object's value.

The value of the object is the data that is stored in it. Data in an object can be stored as numbers, strings, or references to other objects. Strings, and integers are objects themselves. Hence, for objects that are not implemented in C (or core objects), it is a reference to other objects, and we perceive value as the group value of the referenced object. Let's take an example of an object **iC** instance of the C class with the `str` and `lst` attributes, as shown in the following diagram:

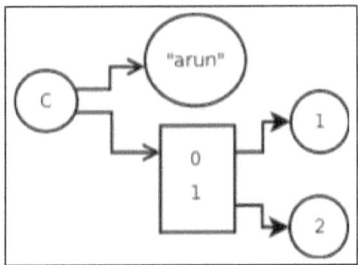

The code snippet to create this object will be as follows:

```
>>> class C:
...     def __init__(self, arg1, arg2):
...         self.str = arg1
...         self.lst = arg2
...
>>> iC = C("arun",[1,2])
>>> iC.str
'arun'
>>> iC.ls
t [1, 2]
>>> iC.lst.append(4)
>>> iC.lst
[1, 2, 4]
```

Then, when we modify **iC**, we are either changing the objects references via attributes, or we are changing the references themselves and not the object **iC**. This is important in understanding immutable objects because being immutable means not being able to change references. Hence, we can change mutable objects that are referenced by immutable objects. For example, lists inside tuple can be changed because the referenced objects are changing, not the references.

Type

Key 3: Type is instance's class.

An object's type tells us about the operations and functionality that the object supports, and it may also define the possible values for objects of that type. For example, your pet may be of type dog (an instance of the dog class) or cat (an instance of the cat class). If it is of type dog, it can bark; and if it is type cat, it can meow. Both are a type of animal (cat and dog inherit from the animal class).

An object's class provides a type to it. Interpreter gets the object's class by checking its __class__ attribute. So, we can change an object's type by changing its __class__ attribute:

```
>>> k = []
>>> k.__class__
_ <class
'list'>
>>> type(k)
<class 'list'>
# type is instance's class
>>> class M:
...     def __init__(self,d):
...         self.d = d
...     def square(self):
...         return self.d * self.d
...
>>>
>>> class N:
...     def __init__(self,d):
...         self.d = d
...     def cube(self):
...         return self.d * self.d * self.d
...
>>>
>>> m = M(4)
>>> type(m) #type is its class
<class '__main__.M'>
```

```
>>> m.square()  #square defined in class M
16
>>> m.__class__ = N # now type should change
>>> m.cube()        # cube in class N
64
>>> type(m)
<class '__main__.N'> # yes type is changed
```

 This will not work for built-in, compiled classes as it works only for class objects defined on runtime.

Making calls to objects

Key 4: All objects can be made callable.

To reuse and group code for some task, we group it in the functions classes, and then call it with different inputs. The objects that have a __call__ attribute are callable and __call__ is the entry point. For the C class, tp_call is checked in its structure:

```
>>> def func(): # a function
...print("adf")
...
>>> func()
adf
>>> func.__call__() #implicit call method
adf
>>> func.__class__.__call__(func)
adf
>>> func.__call__
<method-wrapper '__call__' of function object at 0x7ff7d9f24ea0>
>>> class C: #a callable class
...def __call__(self):
...print("adf")
...
>>> c = C()
>>> c()
adf
>>> c.__call__() #implicit passing of self
adf
>>> c.__class__.__call__(c) #explicit passing of self
adf
>>> callable(lambda x:x+1)  #testing whether object is callable or not
True
```

```
>>> isinstance(lambda x:x+1, collections.Callable) #testing whether
object is callable or not
True
```

Methods in classes are similar to functions, except that they are called with an implicit instance as a first argument. The functions are exposed as methods when they are accessed from the instance. The function is wrapped in a method class and returned. The method class stores instances in __self__ and function in __func__, and its __call__ method calls __func__ with first argument as __self__:

```
>>> class D:
...pass
...
>>> class C:
...      def do(self,):
...              print("do run",self)
...
>>> def doo(obj):
...      print("doo run",obj)
...
>>> c = C()
>>> d = D()
>>> doo(c)
doo run <__main__.C object at 0x7fcf543625c0>
>>> doo(d)
doo run <__main__.D object at 0x7fcf54362400>
>>> # we do not need to pass object in case of C class do method
...
>>> c.do() #implicit pass of c object to do method
do run <__main__.C object at 0x7fcf543625c0>
>>> C.doo = doo
>>> c.doo()
doo run <__main__.C object at
0x7fcf543625c0> >>> C.doo()
Traceback (most recent call last):
  File "<stdin>", line 1, in <module>
TypeError: doo() missing 1 required positional argument: 'obj'
>>> C.do()
Traceback (most recent call last):
  File "<stdin>", line 1, in <module>
TypeError: do() missing 1 required positional argument: 'self'
>>> C.do(c)
```

```
do run <__main__.C object at 0x7fcf543625c0>
>>> C.do(d)
do run <__main__.D object at 0x7fcf54362400>
>>> c.do.__func__(d) #we called real function this way
do run <__main__.D object at 0x7fcf54362400>
```

Using this logic, we can also collect methods that are needed from other classes in the current class, like the following code, instead of multiple inheritances if data attributes do not clash. This will result in two dictionary lookups for an attribute search: one for instance, and one for class.

```
>>> #in library
... class PrintVals:
...     def __init__(self, endl):
...         self.endl = endl
...
...     def print_D8(self, data):
...         print("{0} {1} {2}".format(data[0],data[1],self.endl))
...
>>> class PrintKVals: #from in2 library
...def __init__(self, knm):
...         self.knm = knm
...
...     def print_D8(self, data):
...         print("{2}:{0} {1}".format(data[0],data[1],self.knm))
...
>>> class CollectPrint:
...
...def __init__(self, endl):
...         self.endl = endl
...         self.knm = "[k]"
...
...     print_D8 = PrintVals.print_D8
...     print_D8K = PrintKVals.print_D8
...
>>> c = CollectPrint("}")
>>> c.print_D8([1,2])
1 2 }
>>> c.print_D8K([1,2])
[k]:1 2
```

When we call classes, we are calling its type, that is `metaclass`, with class as a
first argument to give us a new instance:

```
>>> class Meta(type):
...     def __call__(*args):
...         print("meta call",args)
...
>>> class C(metaclass=Meta):
...pass
...
>>> c = C()
meta call (<class '__main__.C'>,)
>>> c = C.__class__.__call__(C)
meta call (<class '__main__.C'>,)
```

Similarly, when we call instances, we are calling their type, that is class, with
instance as first argument:

```
>>> class C:
...     def __call__(*args):
...         print("C call",args)
...
>>> c = C()
>>> c()
C call (<__main__.C object at 0x7f5d70c2bb38>,)
>>> c.__class__.__call__(c)
C call (<__main__.C object at 0x7f5d70c2bb38>,)
```

How objects are created

Objects other than built-in types or compiled module classes are created at runtime.
Objects can be classes, instances, functions, and so on. We call an object's type to
give us a new instance; or put in another way, we call a `type` class to give us an
instance of that type.

Creation of function objects

Key 5: Create function on runtime.

Let's first take a look at how function objects can be created. This will broaden our view. This process is done by interpreter behind the scenes when it sees a `def` keyword. It compiles the code, which is shown as follows, and passes the code name arguments to the function class that returns an object:

```
>>> function_class = (lambda x:x).__class__
>>> function_class
<class 'function'>
>>> def foo():
...     print("hello world")
...
>>>
>>> def myprint(*args,**kwargs):
...print("this is my print")
...print(*args,**kwargs)
...
>>> newfunc1 = function_class(foo.__code__, {'print':myprint})
>>> newfunc1()
this is my print
hello world
>>> newfunc2 = function_class(compile("print('asdf')","filename","single"),{'print':print})
>>> newfunc2()
asdf
```

Creation of instances

Key 6: Process flow for instance creation.

We call class to get a new instance. We saw from the making calls to objects section that when we call class, it calls its metaclass __call__ method to get a new instance. It is the responsibility of __call__ to return a new object that is properly initialized. It is able to call class's __new__, and __init__ because class is passed as first argument, and instance is created by this function itself:

```
>>> class Meta(type):
...     def __call__(*args):
...         print("meta call ",args)
...         return None
...
>>>
>>> class C(metaclass=Meta):
...def __init__(*args):
...         print("C init not called",args)
...
>>> c = C() #init will not get called
meta call (<class '__main__.C'>,)
>>> print(c)
None
>>>
```

To enable developer access to both functionalities, creating new object, and initializing new object, in class itself; __call__ calls the __new__ class to return a new object and __init__ to initialize it. The full flow can be visualized as shown in the following code:

```
>>> class Meta(type):
...     def __call__(*args):
...         print("meta call ,class object :",args)
...         class_object = args[0]
...         if '__new__' in class_object.__dict__:
...             new_method = getattr(class_object,'__new__',None)
...             instance = new_method(class_object)
...         else:
...             instance = object.__new__(class_object)
...         if '__init__' in class_object.__dict__:
...             init_method = getattr(class_object,'__init__',None)
...             init_method(instance,*args[1:])
...         return instance
...
```

```
>>> class C(metaclass=Meta):
...     def __init__(instance_object, *args):
...         print("class init",args)
...     def __new__(*args):
...         print("class new",args)
...         return object.__new__(*args)
...
>>> class D(metaclass=Meta):
...pass
...
>>> c=C(1,2)
meta call ,class object : (<class '__main__.C'>, 1, 2)
class new (<class '__main__.C'>,)
class init (1, 2)
>>> d = D(1,2)
meta call ,class object : (<class '__main__.D'>, 1, 2)
>>>
```

Take a look at the following diagram:

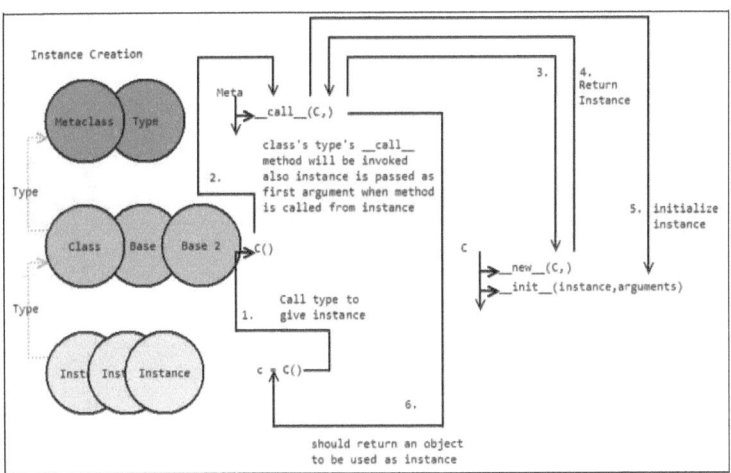

Creation of class objects

Key 7: Process flow for class creation.

There are three ways in which we can create classes. One is to simply define the class. The second one is to use the built-in __build_class__ function, and the third is to use the new_class method of type module. Method one uses two, method two uses method three internally. When interpreter sees a class keyword, it collects the name, bases, and metaclass that is defined for the class. It will call the __build_class__ built-in function with function (with the code object of the class), name of the class, base classes, metaclass that is defined, and so on:

```
__build_class__(func, name, *bases, metaclass=None, **kwds) -> class
```

This function returns the class. This will call the __prepare__ class method of metaclass to get a mapping data structure to use as a namespace. The class body will be evaluated, and local variables will be stored in this mapping data structure. Metaclass's type will be called with this namespace dictionary, bases, and class name. It will in turn call the __new__ and __init__ methods of metaclass. Metaclass can change attributes passed to its method:

```
>>> function_class = (lambda x:x).__class__
>>> M = __build_class__(function_class(
...                     compile("def __init__(self,):\n
print('adf')",
...                             '<stdin>',
...                             'exec'),
...                     {'print':print}
...                     ),
...                 'MyCls')
>>> m = M()
adf
>>> print(M,m)
<class '__main__.MyCls'> <__main__.MyCls object at 0x0088B430>
>>>
```

Take a look at the following diagram:

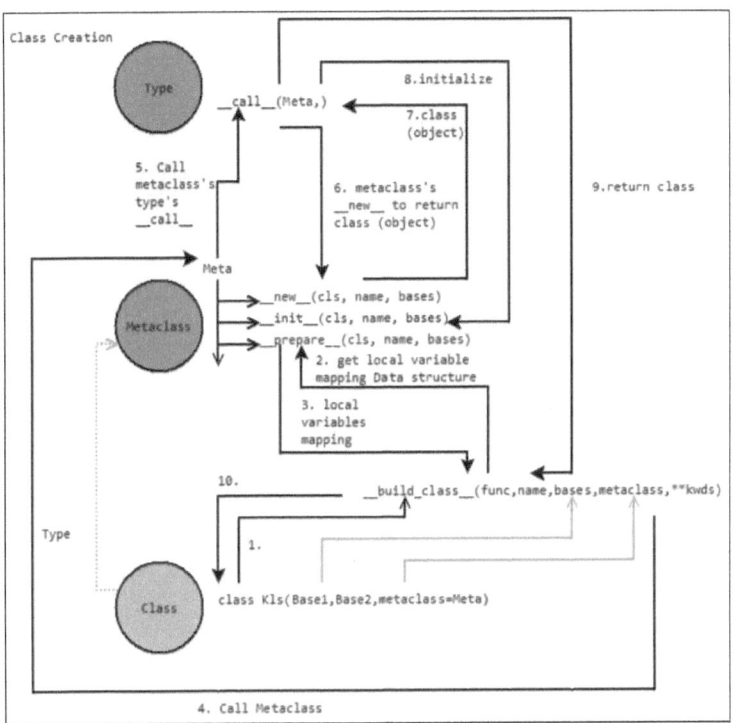

Playing with attributes

Key 8: Which attribute will be used.

Attributes are values that are associated with an object that can be referenced by name using dotted expressions. It is important to understand how attributes of an object are found. The following is the sequence that is used to search an attribute:

1. If an attribute is a special method, and it exists in the object's type (or bases), return it, for example: __call__, __str__, and __init__. When these methods are searched, their behavior is only in the instance's type:

    ```
    >>> class C:
    ...     def __str__(self,):
    ...             return 'Class String'
    ...     def do(self):
    ...             return 'Class method'
    ...
    >>> c = C()
    >>> print(c)
    Class String
    >>> print(c.do())
    Class method
    >>> def strf(*args):
    ...     return 'Instance String',args
    ...
    >>> def doo(*args):
    ...     return 'Instance Method'
    ...
    >>> c.do = doo
    >>> c.__str__ = strf
    >>> print(c)

    >>> print(c.do())
    Instance Method
    ```

2. If an object's type has a __getattribute__ attribute, then this method is invoked to get the attribute whether this attribute is present or not. It is the total responsibility of __getattribute__ to get the attribute. As shown in the following code snippet, even if the do method is present, it is not found as getattribute didn't return any attribute:

    ```
    >>> class C:
    ...     def do(self):
    ...             print("asdf")
    ...     def __getattribute__(self,attr):
    ```

```
    ...              raise AttributeError('object has no attribute
    "%s"'%attr)
    ...
    >>> c = C()
    >>> c.do()
    Traceback (most recent call last):
      File "<stdin>", line 1, in <module>
      File "<stdin>", line 5, in
    __getattribute__ AttributeError: object has
    no attribute "do" >>>
```

3. Search in object's type __dict__ to find the attribute. If it is present, and it is data descriptor, return it:

```
    >>> class Desc:
    ...      def __init__(self, i):
    ...              self.i = i
    ...      def __get__(self, obj, objtype):
    ...              return self.i
    ...      def __set__(self,obj, value):
    ...              self.i = value
    ...
    >>> class C:
    ...      attx = Desc(23)
    ...
    >>> c = C()
    >>> c.attx
    23
    >>> c.__dict__['attx'] = 1234
    >>> c.attx
    23
    >>> C.attx = 12
    >>> c.attx
    1234
```

4. Search in object's __dict__ type (and if this object is class, search bases __dict__ as well) to find the attribute. If the attribute is descriptor, return the result.

5. Search in object's type __dict__ to find the attribute. If the attribute is found, return it. If it is non-data descriptor, return its result, and check in other bases using the same logic:

```
    >>> class Desc:
    ...      def __init__(self, i):
    ...              self.i = i
```

```
...        def __get__(self, obj, objtype):
...                return self.i
...
>>> class C:
...        attx = Desc(23)
...
>>> c = C()
>>> c.attx
23
>>> c.__dict__['attx'] = 34
>>> c.attx
34
```

6. If object type's __getattr__ is defined, check whether it can give us the attribute:

```
>>> class C:
...        def __getattr__(self, key):
...                return key+'_#'
...
>>> c = C()
>>> c.asdf
'asdf_#'
```

7. Raise AttributeError.

Descriptors

Key 9: Making custom behavior attributes.

Any attribute of a class, which is an object defining any of these methods, acts as a descriptor:

- __get__(self, obj, type=None) --> value
- __set__(self, obj, value) --> None
- __delete__(self, obj) --> None

When an attribute is searched in an object first, it is searched in its dictionary then its type's (base class's) dictionary. If found, object has one of these methods defined and that method is invoked instead. Let's assume that b is an instance of the B class, then the following will happen:

- Invocation through class is type.__getattribute__() transforming to B.__dict__['x'].__get__(None, B)
- Invocation through instance is object.__getattribute__() --> type(b).__dict__['x'].__get__(b, type(b))

Objects with only __get__ are non-data descriptors, and objects that include __set__ / __del__ are data descriptors. Data descriptors take precedence over instance attributes, whereas non-data descriptors do not.

Class, static, and instance methods

Key 10: Implementing class method and static method.

Class, static, and instance methods are all implementable using descriptors. We can understand descriptors and these methods in one go:

- Class methods are methods that always get class as their first argument and they can be executed without any instance of class.
- Static methods are methods that do not get any implicit objects as first argument when executed via class or instance.
- Instance methods get instances when called via instance but no implicit argument when called via class.

A sample code usage of these methods is as follows:

```
>>> class C:
...     @staticmethod
...     def sdo(*args):
...             print(args)
...     @classmethod
...     def cdo(*args):
...             print(args)
...     def do(*args):
...             print(args)
...
>>> ic = C()
# staticmethod called through class: no implicit argument is passed
>>> C.sdo(1,2)
(1, 2)
# staticmethod called through instance:no implicit argument is passed
>>> ic.sdo(1,2)
(1, 2)
# classmethod called through instance: first argument implicitly class
>>> ic.cdo(1,2)
(<class '__main__.C'>, 1, 2)
# classmethod called through class: first argument implicitly class
>>> C.cdo(1,2)
```

```
(<class '__main__.C'>, 1, 2)
# instancemethod called through instance: first argument implicitly
instance
>>> ic.do(1,2)
(<__main__.C object at 0x00DC9E30>, 1, 2)
#instancemethod called through class: no implicit argument, acts like
static method.
>>> C.do(1,2)
(1, 2)
```

They can be understood and implemented using descriptors easily as follows:

```
from functools import partial
>>> class my_instancemethod:
...def __init__(self, f):
...            # we store reference to function in instance
...            # for future reference
...            self.f = f
...      def __get__(self, obj, objtype):
...            # obj is None when called from class
...            # objtype is always present
...            if obj is not None:
...                return partial(self.f,obj)
...            else: # called from class
...                return self.f
...
>>> class my_classmethod:
...      def __init__(self, f):
...            self.f = f
...      def __get__(self, obj, objtype):
...            # we pass objtype i.e class object
...            # when called from instance or class
...            return partial(self.f,objtype)
...
>>> class my_staticmethod:
...def __init__(self, f):
...            self.f = f
...      def __get__(self, obj, objtype):
...            # we do not pass anything
...            # for both conditions
...            return self.f
...
>>> class C:
...      @my_instancemethod
...      def ido(*args):
```

```
...             print("imethod",args)
...         @my_classmethod
...         def cdo(*args):
...             print("cmethod",args)
...         @my_staticmethod
...         def sdo(*args):
...             print("smethod",args)
...
>>> c = C()
>>> c.ido(1,2)
imethod (<__main__.C object at 0x00D7CBD0>, 1, 2)
>>> C.ido(1,2)
imethod (1, 2)
>>> c.cdo(1,2)
cmethod (<class '__main__.C'>, 1, 2)
>>> C.cdo(1,2)
cmethod (<class '__main__.C'>, 1, 2)
>>> c.sdo(1,2)
smethod (1, 2)
>>> C.sdo(1,2)
smethod (1, 2)
```

Summary

In this chapter, we dived into how objects work in the Python language, how are they connected, and how are they called. Descriptors and instance creation are very important topics as they give us a picture of how system works. We also dived into how attributes are looked up for objects.

Now, we are all prepared to learn how to use language constructs to their maximum potential. In the next chapter, we will also discover utilities that are extremely helpful in elegantly finishing a project.

2

Namespaces and Classes

In the previous chapter, we covered how objects worked. In this chapter, we will explore how objects are made available to code via reference, specifically how namespaces work, what modules are, and how they are imported. We will also cover topics related to classes, such as language protocols, MRO, and abstract classes. We will discuss the following topics:

- Namespaces
- Imports and modules
- Class multiple inheritance, MRO, super
- Protocols
- Abstract classes

How referencing objects work
– namespaces

Key 1: Interrelations between objects.

The scope is the visibility of a name within a code block. Namespace is mapping from names to objects. Namespaces are important in order to maintain localization and avoid name collision. Every module has a global namespace. Modules store mapping from variable name to objects in their `__dict__` attribute, which is a normal Python dictionary along with information to reload it, package information, and so on.

Every module's global namespace has an implicit reference to the built-in module; hence, objects that are in the built-in module are always available. We can also import other modules in the main script. When we use the syntax import module name, a mapping with module name to module object is created in the global namespace of the current module. For import statements with syntax such as `import modname as modrename`, mapping is created with a new name to module object.

We are always in the __main__ module's global namespace when the program starts, as it is the module that imports all others. When we import a variable from another module, only an entry is created for that variable in the global namespace pointing at the referenced object. Now interestingly, if this variable references a function object, and if this function uses a global variable, then this variable will be searched in the global namespace of the module that the function was defined in, not in the module that we imported this function to. This is possible because functions have the __globals__ attribute that points to its __dict__ modules, or in short, its modules namespace.

All modules that are loaded and referenced are cached in `sys.modules`. All imported modules are names pointing to objects in `sys.modules`. Let's define a new module like this with the name `new.py`:

```
k = 10
def foo():
    print(k)
```

By importing this module in the interactive session, we can see how global namespaces work. When this module is reloaded, its namespace dictionary is updated, not recreated. Hence, if you attach anything new from the outside of the module to it, it will survive reload:

```
>>> import importlib
>>> import new
>>> from new import foo
>>> import sys
>>> foo()
10
>>> new.foo()
10
>>> foo.__globals__ is sys.modules['new'].__dict__  # dictionary
used by namespace and function attribute __globals__ is indeed same
True
>>> foo.__globals__['k'] = 20  # changing global namespace dictionary
>>> new.do   #attribute is not defined in the module
Traceback (most recent call last):
  File "<stdin>", line 1, in <module>
```

```
AttributeError: module 'new' has no attribute 'do'
>>> foo.__globals__['do'] = 22 #we are attaching attribute to
module from outside the module
>>> new.do
22
>>> foo()  # we get updated value for global variable
20
>>> new.foo()
20
>>> importlib.reload(new) #reload repopulates old modules
dictionary <module 'new' from '/tmp/new.py'>
>>> new.do #it didn't got updated as it was populated from outside.
22
>>> new.foo() #variables updated by execution of code in module are
updated
10
```
>>>

If we use the functions that are defined in different modules to compose a class on runtime, such as using metaclasses, or class decorators, this can bring up surprises as each function could be using a different global namespace.

Locals are simple and they work in the way that you expect. Each function call gets its own copy of variables. Nonlocal variables make variables that are defined in the outer scope (not global namespace) accessible to the current code block. In the following code example, we can see how variables can be referenced in enclosed functions.

Code blocks are able to reference variables that are defined in enclosing scopes. Hence, if a variable is not defined in a function but in an enclosing function, we are able to get its value. If, after referencing a variable in an outer scope, we assign a value to this variable in a code block, it will confuse the interpreter in finding the right variable, and we will get the value from the current local scope. If we assign a value to the variable, it defaults to the local variable. We can specify that we want to work with an enclosing variable using a nonlocal keyword:

```
>>> #variable in enclosing scope can be referenced any level deep
...
>>> def f1():
...     v1 = "ohm"
...     def f2():
...         print("f2",v1)
...         def f3():
...             print("f3",v1)
...         f3()
...     f2()
```

```
    ...
    >>> f1()
    f2 ohm
    f3 ohm
>>>
    >>> #variable can be made non-local (variable in outer scopes)
    skipping one level of enclosing scope
    ...
    >>> def f1():
    ...     v1 = "ohm"
    ...     def f2():
    ...         print("f2",v1)
    ...         def f3():
    ...             nonlocal v1
    ...             v1 = "mho"
    ...             print("f3",v1)
    ...         f3()
    ...         print("f2",v1)
    ...     f2()
    ...     print("f1",v1)
    ...
    >>> f1()
    f2 ohm
    f3 mho
    f2 mho
    f1 mho
>>>
>>>
    >>> #global can be specified at any level of enclosed function
    ...
    >>> v2 = "joule"
    >>>
    >>> def f1():
    ...     def f2():
    ...         def f3():
    ...             global v2
    ...             v2 = "mho"
    ...             print("f3",v2)
    ...         f3()
    ...         print("f2",v2)
    ...     f2()
    ...     print("f1",v2)
    ...
    >>> f1()
```

```
f3 mho
f2 mho
f1 mho
```

As variables are searched without any dictionary lookup for the local namespace, it is faster to look up variables inside a function with a small number of variables than to search in a global namespace. On similar lines, we will get a little speed boost if we pull objects that are referenced in loops in a function's local namespace inside a function block:

```
In [6]: def fun():
   ...:        localsum = sum
   ...:        return localsum(localsum((a,a+1)) for a in range(1000))
   ...:

In [8]: def fun2():
   ...:        return sum(sum((a,a+1)) for a in range(1000))
   ...:

In [9]: %timeit fun2()
1000 loops, best of 3: 1.07 ms per loop

In [11]: %timeit fun()
1000 loops, best of 3: 983 µs per loop
```

Functions with state – closures

Key 2: Creating cheap state-remembering functions.

A closure is a function that has access to variables in an enclosing scope, which has completed its execution. This means that referenced objects are kept alive until the function is in memory. The main utility of such a setup is to easily retain some state, or to create specialized functions whose functioning depends on the initial setup:

```
>>> def getformatter(start,end):
...def formatter(istr):
...print("%s%s%s"%(start,istr,end))
...return formatter
...
>>> formatter1 = getformatter("<",">")
>>> formatter2 = getformatter("[","]")
>>>
>>> formatter1("hello")
<hello>
>>> formatter2("hello")
```

```
[hello]
>>> formatter1.__closure__[0].cell_contents
'>'
>>> formatter1.__closure__[1].cell_contents
'<'
```

We can do the same by creating a class and using the instance object to save state. The benefit with closures is that variables are stored in a __closure__ tuple, and hence, they are fast to access. Less code is required to create a closure as compared to classes:

```
>>> def formatter(st,en):
...def fmt(inp):
...              return "%s%s%s"%(st,inp,en)
...     return fmt
...
>>> fmt1 = formatter("<",">")
>>> fmt1("hello")
'<hello>'
>>> timeit.timeit(stmt="fmt1('hello')",
... number=1000000,globals={'fmt1':fmt1})
0.3326794120075647
>>> class Formatter:
...     def __init__(self,st,en):
...             self.st = st
...             self.en = en
...     def __call__(self, inp):
...             return "%s%s%s"%(self.st,inp,self.en)
...
>>> fmt2 = Formatter("<",">")
>>> fmt2("hello")
'<hello>'
>>> timeit.timeit(stmt="fmt2('hello')",
... number=1000000,globals={'fmt2':fmt2})
0.5502702980011236
```

One such function is available from the standard library, named partial, that makes use of closure to create a new function that is always invoked with some predefined arguments:

```
>>> import functools
```

>>>

```
>>> def foo(*args,**kwargs):
...     print("foo with",args,kwargs)
...
```

```
>>> pfoo = functools.partial(foo,10,20,v1=23)
>>> foo(1,2,3,array=1)
foo with (1, 2, 3) {'array': 1}
>>> pfoo()
foo with (10, 20) {'v1': 23}
>>> pfoo(30,40,array=12)
foo with (10, 20, 30, 40) {'v1': 23, 'array': 12}
```

Understanding import and modules

Key 3: Creating a custom loader for modules.

Import statements get references of other module objects in the current module's namespace. It consists of searching the module, executing code to create a module object, updating caches (sys.modules), updating modules namespace, and creating a reference to new module being imported.

The built-in __import__ function searches and executes the module to create a module object. The importlib library has the implementation, and it also provides a customizable interface to the import mechanism. Various classes interact to get the job done. The __import__ function should return a module object. For example, in the following example, we are creating a module finder, which checks for modules in any path that is given as an argument during construction. Here, an empty file named names.py should be present at the given path. We have loaded the module, then inserted its module object in sys.modules and added a function to this module's global namespace:

```
import os
import sys

class Spec:
    def __init__(self,name,loader,file='None',path=None,
                    cached=None,parent=None,has_location=False):
        self.name = name
        self.loader = loader
        self.origin = file
        self.submodule_search_locations = path
        self.cached = cached
        self.has_location = has_location


class Finder:
```

```
    def __init__(self, path):
        self.path = path

    def find_spec(self,name,path,target):
        print("find spec name:%s path:%s
target:%s"%(name,path,target))
        return Spec(name,self,path)

    def load_module(self, fullname):
        print("loading module",fullname)
        if fullname+'.py' in os.listdir(self.path):
            import builtins
            mod = type(os)
            modobject = mod(fullname)
            modobject.__builtins__ = builtins
            def foo():
                print("hii i am foo")
            modobject.__dict__['too'] = foo
            sys.modules[fullname] = modobject
            modobject.__spec__ = 'asdfasfsadfsd'
            modobject.__name__ = fullname
            modobject.__file__ = 'aruns file'
            return modobject

sys.meta_path.append(Finder(r'/tmp'))
import notes
notes.too()

Output:
find spec name:notes path:None target:None
loading module notes
hii i am foo
```

Customizing imports

If the module has an __all__ attribute, only the names that are specified by the iterable in this attribute will be imported from module import *. Let's assume that we created a module named mymod.py, as follows:

```
__all__ = ('hulk','k')

k = 10
def hulk():
```

```
    print("i am hulk")

def spidey():
    print("i am spidey")
```

We will not be able to import `spidey` from `mymod` as it is not included in `__all__`:

```
>>> from mymod import *
```
>>>
```
>>> hulk()
i am hulk
>>> k
10
>>> spidey()
Traceback (most recent call last):
  File "<stdin>", line 1, in <module>
NameError: name 'spidey' is not defined
```

Class inheritance

We already discussed how instances and classes are created. We also discussed how attributes are accessed in a class. Let's dive deeper into how this works for multiple base classes. As type is searched for the presence of an attribute for an instance, if the type inherits from a number of classes, they all are searched as well. There is a defined pattern to this (**Method Resolution Order (MRO)**). This order plays an important role in determining the method in cases of multiple inheritance and diamond-shaped inheritance.

Method resolution order

Key 4: Understanding MRO.

The methods are searched in the base classes of a class in predefined manner. This sequence or order is known as method resolution order. In Python 3, when an attribute is not found in a class, it is searched in all the base classes of that class. If the attribute is still not found, the base classes of the base classes are searched. This process goes on until we exhaust all base classes. This is similar to how if we have to ask a question, we will first go to our parents and then to uncles, and aunts (the same level base classes). If we still do not get an answer, we will approach grandparents. The following code snippet shows this sequence:

```
>>> class GrandParent:
...def do(self,):
```

```
...         print("Grandparent do called")
...
>>> class Father(GrandParent):
...def do(self,):
...         print("Father do called")
...
>>> class Mother(GrandParent):
...def do(self,):
...         print("Mother do called")
...
>>> class Child(Father, Mother):
...def do(self,):
...         print("Child do called")
...
>>> c = Child() # calls method in Class
>>> c.do()
Child do called
>>> del Child.do # if method is not defined it is searched in bases
>>> c.do()   #Father's method
Father do called
>>> c.__class__.__bases__ = (c.__class__.__bases__[1],c.__class__.__
bases__[0]) #we swap bases order
>>> c.do() #Mothers's method
Mother do called
>>> del Mother.do
>>> c.do() #Fathers' method
Father do called
>>> del Father.do
>>> c.do()
Grandparent do called
```

Super's superpowers

Key 6: Get superclass's methods without a superclass definition.

We mostly create subclasses to specialize methods or add a new functionality. We may need to add some feature, which is 80% the same as one in the base class. Then it will be natural to call base class's method for that portion of functionality and add extra functionality in new method in the subclass. To call a method in superclass, we can either use its class name to access the method, or super it like this:

```
>>> class GrandParent:
...def do(self,):
```

```
...          print("Grandparent do called")
...
>>> class Father(GrandParent):
...def do(self,):
...          print("Father do called")
...
>>> class Mother(GrandParent):
...def do(self,):
...          print("Mother do called")
...
>>> class Child(Father, Mother):
...def do(self,):
...          print("Child do called")
...
>>> c = Child()
>>> c.do()
Child do called
>>> class Child(Father, Mother):
...def do(self,):
...          print("Child do called")
...          super().do()
...
>>> c = Child()
>>> c.do() Child
do called Father
do called
>>> print("Father and child super
calling") Father and child super calling
>>> class Father(GrandParent):
...     def do(self,):
...          print("Father do called")
...          super().do()
...
>>> class Child(Father, Mother):
...def do(self,):
...          print("Child do called")
...          super().do()
...
>>> c = Child()
>>> c.do() Child
do called Father
do called Mother
do called
```

```
>>> print("Father and Mother super calling")
Father and Mother super calling
>>> class Mother(GrandParent):
...     def do(self,):
...             print("Mother do called")
...             super().do()
...
>>> class Father(GrandParent):
...def do(self,):
...             print("Father do called")
...             super().do()
...
>>> class Child(Father, Mother):
...def do(self,):
...             print("Child do called")
...             super().do()
...
>>> c = Child()
>>> c.do() Child do
called Father do
called Mother do
called Grandparent
do called
>>> print(Child.__mro__)
(<class '__main__.Child'>, <class '__main__.Father'>, <class '__
main__.Mother'>, <class '__main__.GrandParent'>, <class 'object'>)
```

Using language protocols in classes

All objects that provide a specific functionality have certain methods that facilitate that behavior, for example, you can create an object of type worker and expect it to have the `submit_work(function, kwargs)`, and `is_completed()` methods. Now, we can expect all objects that have these methods to be usable as workers in any application portion. Similarly, the Python language has defined some methods that are needed to add a certain functionality to an object. If an object possesses these methods, it has that functionality.

We will discuss two very import protocols: iteration protocol, and context protocol.

Iteration protocol

For iteration protocol, objects must possess the __iter__ method. If the object possesses it, we can use the object anywhere that we use an iterator object. When we are using the iterator object in a `for` loop or passing it to the `iter` built-in function, we are calling its __iter__ method. This method returns another or the same object that is responsible for maintaining the index during iteration, and this object that is returned from __iter__ must have a __next__ method that provides the next values in sequence and raises `StopIteration` on the finish of this sequence. In the following code snippet, the `BooksIterState` objects help retain the index that is used for iteration. If the books __iter__ method returned self, then it would be difficult to maintain a state index when the object is accessed from two loops:

```
>>> class BooksIterState:
...     def __init__(self, books):
...             self.books = books
...             self.index = 0
...     def __next__(self,):
...             if self.index >= len(self.books._data):
...                     raise StopIteration
...             else:
...                     tmp = self.books._data[self.index]
...                     self.index += 1
...                     return tmp
...
>>> class Books:
...     def __init__(self, data):
...             self._data = data
...     def __iter__(self,):
...             return BooksIterState(self)
...
>>> ii = iter(Books(["don quixote","lord of the
flies","great expectations"]))
>>> next(ii)
'don quixote'
>>> for i in Books(["don quixote","lord of the
flies","great expectations"]):
...print(i)
...
don quixote
lord of the flies
great expectations
>>> next(ii)
'lord of the flies'
>>> next(ii)
```

```
'great expectations'
>>> next(ii)
Traceback (most recent call last):
  File "<stdin>", line 1, in <module>
  File "<stdin>", line 7, in __next__
StopIteration
>>>
```

Context manager protocol

The objects providing context for execution are like try finally statements. If an object has the __enter__ and __exit__ methods, then this object can be used as a replacement of try finally statements. The most common uses are releasing locks and resources, or flushing and closing files. In the following example, we are creating a Ctx class to serve as context manager:

```
>>> class Ctx:
...     def __enter__(*args):
...         print("entering")
...         return "do some work"
...     def __exit__(self, exception_type,
...                     exception_value,
...                     exception_traceback):
...         print("exit")
...         if exception_type is not None:
...             print("error",exception_type)
...         return True
...
>>> with Ctx() as k:
...print(k)
...raise KeyError
...
entering
do some work
exit
error <class 'KeyError'>
```

We can also use the contextmanager decorator of contextlib to easily create context managers like the one shown in the following code:

```
>>> import contextlib
>>> @contextlib.contextmanager
... def ctx():
...try:
```

```
...          print("start")
...          yield "so some work"
...      except KeyError:
...          print("error")
...      print("done")
...
>>> with ctx() as k:
...print(k)
...raise KeyError
...
start
so some work
error
done
```

There are other methods that one should know, such as __str__, __add__, __ getitem__, and so on, that define various functionalities of the objects. There is a list of them at the language reference's datamodel.html. You should at least read it once to get to know what methods are available. Here is the link: https://docs.python. org/3/reference/datamodel.html#special-method-names.

Using abstract classes

Key 6: Making interfaces for conformity.

Abstract classes are available via the standard abc library package. They are useful for the definition of interfaces and common functionality. These abstract classes can implement a portion of the interface and make the rest of the API mandatory for subclasses by defining their methods as abstract. Also, classes can be turned into subclasses of the abstract class by simply registering them. These classes are useful to make a set of classes conform to a single interface. Here is how to use them. Here, worker class defines an interface with two methods, do and is_busy, which each type of worker must implement. ApiWorker is the implementation for this interface:

```
>>> from abc import ABCMeta, abstractmethod
>>> class Worker(metaclass=ABCMeta):
...      @abstractmethod
...      def do(self, func, args, kwargs):
...          """ work on function """
...      @abstractmethod
...      def is_busy(self,):
...          """ tell if busy """
...
>>> class ApiWorker(Worker):
```

```
...     def __init__(self,):
...         self._busy = False
...     def do(self, func, args=[], kwargs={}):
...         self._busy = True
...         res = func(*args, **kwargs)
...         self._busy = False
...         return res
...     def is_busy(self,):
...         return self._busy
...
>>> apiworker = ApiWorker()
>>> print(apiworker.do(lambda x: x + 1, (1,)))
2
>>> print(apiworker.is_busy())
False
```

Summary

Now, we have seen how to manipulate namespaces, and to create custom module-loading classes. We can use multiple inheritance to create mixin classes in which each mixin class provides a new functionality to the subclass. Context manager and iterator protocols are very useful constructs to create clean code. We created abstract classes that can help us in setting up API contracts for classes.

In the next chapter, we will cover the functions and utilities that are available to us from a standard Python installation.

3
Functions and Utilities

After learning about how objects are linked to one another, let's take a look at the functions that are the means to execute code in language. We will discuss how to define and call functions with various combinations. Then, we will cover some very useful utilities that are available to us to use in day-to-day programming. We will cover the following topics:

- Defining functions
- Decorating callables
- Utilities

Defining functions

Key 1: How to define functions.

Functions are used to group a set of instructions and logic that performs a specific task. So, we should make functions perform one specific task and choose a name that gives us a hint about that. If a function is important and performs complex stuff, we should always add docstrings to this function so that it is easy for us to later visit and modify this function.

While defining a function, we can define the following:

1. Positional arguments (simply pass the object according to position), which are as follows:
```
>>> def foo(a,b):
...    print(a,b)
...
>>> foo(1,2)
1 2
```

2. Default arguments (if value is not passed, the default is used), which are as follows:

```
>>> def foo(a,b=3):
...print(a,b)
...
>>> foo(3)
3 3
>>> foo(3,4)
3 4
```

3. Keyword only arguments (must be passed as a positional or as a keyword argument), which are as follows:

```
>>> def  foo(a,*,b):
...    print(a,b)
...
>>> foo(2,3)
Traceback (most recent call last):
  File "<stdin>", line 1, in <module>
TypeError: foo() takes 1 positional argument but 2 were given
>>> foo(1,b=4)
1 4
```

4. An argument list, which is as follows:

```
>>> def foo(a,*pa):
...    print(a,pa)
...
>>> foo(1)
1 ()
>>> foo(1,2)
1 (2,)
>>> foo(1,2,3)
1 (2,  3)
```

5. A keyword argument dictionary, which is as follows:

```
>>> def foo(a,**kw):
...    print(a,kw)
...
>>> foo(2)
2 {}
>>> foo(2,b=4)
2 {'b': 4}
>>> foo(2,b=4,v=5)
2 {'b': 4, 'v': 5}
```

When a function is called, this is how arguments are passed on:

6. All positional arguments that are passed are consumed.

7. If the function takes an argument list and there are more passed positional arguments after the first step, then the rest of the arguments are collected in an argument list:

```
>>> def foo1(a,*args):
...    print(a,args)
...
>>> def foo2(a,):
...    print(a)
...
>>> foo1(1,2,3,4)
1 (2, 3, 4)
>>> foo2(1,2,3,4)
Traceback (most recent call last):
  File "<stdin>", line 1, in <module>
TypeError: foo2() takes 1 positional argument but 4 were given
```

8. If passed position arguments are less than the defined positional arguments, then the passed keyword arguments are used for values for positional arguments. If no keyword argument is found for the positional argument, we get an error:

```
>>> def foo(a,b,c):
...    print(a,b,c)
...
>>> foo(1,c=3,b=2)
1 2 3
>>> foo(1,b=2)
Traceback (most recent call last):
  File "<stdin>", line 1, in <module>
TypeError: foo() missing 1 required positional argument: 'c'
```

9. Passed keyword variables are used only for keyword arguments:

```
>>> def foo(a,b,*,c):
...    print(a,b,c)
...
>>> foo(1,2,3)
Traceback (most recent call last):
  File "<stdin>", line 1, in <module>
TypeError: foo() takes 2 positional arguments but 3 were given
>>> foo(1,2,c=3)
1 2 3
>>> foo(c=3,b=2,a=1
) 1 2 3
```

10. If more keywords remain and the called function takes a keyword argument list, then the rest of the keyword arguments are passed as a keyword argument list. If the keyword argument list is not taken by the function, we get an error:

```
>>> def foo(a,b,*args,c,**kwargs):
...     print(a,b,args,c,kwargs)
...
>>> foo(1,2,3,4,5,c=6,d=7,e=8)
1 2 (3, 4, 5) 6 {'d': 7, 'e': 8}
```

Here is an example function that uses all of the preceding combinations:

```
>>> def foo(a,b,c=2,*pa,d,e=5,**ka):
...     print(a,b,c,d,e,pa,ka)
...
>>> foo(1,2,d=4)
1 2 2 4 5 () {}
>>> foo(1,2,3,4,5,d=6,e=7,g=10,h=11)
1 2 3 6 7 (4, 5) {'h': 11, 'g': 10}
```

Decorating callables

Key 2: Changing the behavior of callables.

Decorators are callable objects, which replace the original callable objects with some other objects. In this case, as we are replacing a callable with another object, what we mostly want mostly is the replaced object to be callable.

Language provides syntax to do so easily, but first, let's take a look at how we can manually do this:

```
>>> def wrap(func):
...     def newfunc(*args):
...         print("newfunc",args)
...     return newfunc
...
>>> def realfunc(*args):
...     print("real func",args)
...
>>>
>>> realfunc = wrap(realfunc)
>>>
>>> realfunc(1,2,4)
('newfunc', (1, 2, 4))
```

With the decorator syntax, it becomes easy. Taking the definition of wrap and `newfunc` from the preceding code snippet, we get this:

```
>>> @wrap
... def realfunc(args):
...     print("real func",args)
...
>>> realfunc(1,2,4)
('newfunc', (1, 2, 4))
```

To store some kind of state in the decorator function, say to make decorator more useful and applicable to wider application code base, we can use closures or class instances as decorators. In the second chapter, we saw that closures can be used to store state; let's look at how we can utilize them to store information in decorators. In this snippet, the `deco` function is the new function that will replace the add function. A prefix variable is available in the closure of this function. This variable can be injected at decorator creation time:

```
>>> def closure_deco(prefix):
...def deco(func):
...         return lambda x:x+prefix
...     return deco
...
>>> @closure_deco(2)
... def add(a):
...return a+1
...
>>> add(2)
4
>>> add(3)
5
>>> @closure_deco(3)
... def add(a):
...return a+1
...
>>> add(2)
5
>>> add(3)
6
```

We could have used a class to do the same thing as well. Here, we save state on an instance of class:

```
>>> class Deco:
...     def __init__(self,addval):
```

```
...             self.addval = addval
...         def __call__(self, func):
...             return lambda x:x+self.addval
...
>>> @Deco(2)
... def add(a):
...return a+1
...
>>> add(1)
3
>>> add(2)
4
>>> @Deco(3)
... def add(a):
...return a+1
...
>>> add(1)
4
>>> add(2)
5
```

As decorator works on any callable, it works similarly on methods and class definitions as well, but when doing so, we should take into consideration the different arguments that are implicitly passed for the method that is being decorated. Let's first take a simple method being decorated like this:

```
>>> class K:
...     def do(*args):
...         print("imethod",args)
...
>>> k = K()
>>> k.do(1,2,3)
('imethod', (<__main__.K instance at 0x7f12ea070bd8>, 1, 2, 3))
>>>
>>> # using a decorator on methods give similar results
...
>>> class K:
...     @wrap
...     def do(*args):
...         print("imethod",args)
...
>>> k = K()
>>> k.do(1,2,3)
('newfunc', (<__main__.K instance at 0x7f12ea070b48>, 1, 2, 3))
```

As the function that is replaced becomes the method of the class itself, this works perfectly. This is not true for static and class methods. They employ descriptors to call methods, hence, their behavior breaks with decorators and the returned function behaves like a simple method. We can make this work by first checking whether the overridden function is a descriptor and if yes, then calling its __get__ method instead:

```
>>> class K:
...@wrap
...@staticmethod
...def do(*args):
...print("imethod",args)
...@wrap
...@classmethod
...def do2(*args):
...print("imethod",args)
...
>>> k = K()
>>> k.do(1,2,3)
('newfunc', (<__main__.K instance at 0x7f12ea070cb0>, 1, 2, 3))
>>> k.do2(1,2,3)
('newfunc', (<__main__.K instance at 0x7f12ea070cb0>, 1, 2, 3))
```

We can also make this work easily using static and class methods decorators on top of any other decorator. This makes the actual method that is found by the attribute look up as a descriptor and normal execution happens for staticmethod and classmethod.

This works fine, as follows:

```
>>> class K:
...        @staticmethod
...        @wrap
...        def do(*args):
...            print("imethod",args)
...        @classmethod
...        @wrap
...        def do2(*args):
...            print("imethod",args)
...
>>> k = K()
>>> k.do(1,2,3)
('newfunc', (1, 2, 3))
>>> k.do2(1,2,3)
('newfunc', (<class __main__.K at 0x7f12ea05e1f0>, 1, 2, 3))
```

We can use decorators for classes as class is just a type of callable. Hence, we can use decorators to alter the instance creation process so that when we call class, we get an instance. A class object will be passed to decorator and then decorator can replace it with another callable or class. Here, the cdeco decorator is passing a new class that replaced cls:

```
>>> def cdeco(cls):
...     print("cdecorator working")
...     class NCls:
...         def do(*args):
...             print("Ncls do",args)
...     return NCls
...
>>> @cdeco
... class Cls:
...     def do(*args):
...         print("Cls do",args)
...
cdecorator working
>>> b = Cls()
>>> c = Cls()
>>> c.do(1,2,3)
('Ncls do', (<__main__.NCls instance at 0x7f12ea070cf8>, 1, 2, 3))
```

Normally, we use this to change the attributes, and add new attributes to the class definition. We can also use this to register the class to some registry, and so on. In the following code snippet, we check whether class has a do method. If we find one, we replace it with newfunc:

```
>>> def cdeco(cls):
...     if hasattr(cls,'do'):
...         cls.do = wrap(cls.do)
...     return cls
...
>>> @cdeco
... class Cls:
...     def do(*args):
...         print("Cls do",args)
...
>>> c = Cls()
>>> c.do(1,2,3)
('newfunc', (<__main__.Cls instance at 0x7f12ea070cb0>, 1, 2, 3))
```

Utilities

Key 3: Easy iterations by comprehensions.

We have various syntax and utilities to iterate efficiently over iterators. Comprehensions work on iterator and provide results as another iterator. They are implemented in native C, and hence, they are faster than for loops.

We have list, dictionary, and set comprehensions, which produce list, dictionary, and set as result, respectively. Also, iterators avoid declaring extra variables that we need in a loop:

```
>>> ll = [ i+1 for i in range(10)]
>>> print(type(ll),ll)
<class 'list'> [1, 2, 3, 4, 5, 6, 7, 8, 9, 10]
>>> ld = { i:'val'+str(i) for i in range(10) }
>>> print(type(ld),ld)
<class 'dict'> {0: 'val0', 1: 'val1', 2: 'val2', 3: 'val3', 4: 'val4',
5: 'val5', 6: 'val6', 7: 'val7', 8: 'val8', 9: 'val9'}
>>> ls = {i for i in range(10)}
>>> print(type(ls),ls)
<class 'set'> {0, 1, 2, 3, 4, 5, 6, 7, 8, 9}
```

Generator expression creates generators, which can be used to produce generators for an iteration like this. To materialize a generator, we use it to create set, dict, or list:

```
>>> list(( i for i in range(10)))
[0, 1, 2, 3, 4, 5, 6, 7, 8, 9]
>>> dict(( (i,'val'+str(i)) for i in range(10)))
{0: 'val0', 1: 'val1', 2: 'val2', 3: 'val3', 4: 'val4', 5: 'val5', 6:
'val6', 7: 'val7', 8: 'val8', 9: 'val9'}
>>> set(( i for i in range(10)))
{0, 1, 2, 3, 4, 5, 6, 7, 8, 9}
```

Generator objects do not compute all the values of the iterable at once but one by one when requested by a loop. This conserves memory, and we may not be interested in using the whole iterable. Generators are not silver bullets to be used everywhere. They do not always result in a performance increase. It depends on the consumer, and the cost of generating one sequence:

```
>>> def func(val):
...     for i in (j for j in range(1000)):
...         k = i + 5
...
>>> def func_iter(val):
```

```
...        for i in [ j for j in range(1000)]:
...            k = i + 5
...
>>> timeit.timeit(stmt="func(1000)",
globals={'func':func_ iter},number=10000)
0.6765081569974427
>>> timeit.timeit(stmt="func(1000)",
globals={'func':func},numb er=10000)
0.838760247999744
```

Key 4: Some helpful utilities.

The `itertools` utility is a good module with many helpful functions for iterations. Some of my favorites are the following:

- **itertools.chain(* iterable)**: This returns a single iterable from a list of iterables. First, all the elements of the first iterable are exhausted, and then of the second, and so on until all iterables are exhausted:

  ```
  >>> list(itertools.chain(range(3),range(2),range(4)))
  [0, 1, 2, 0, 1, 0, 1, 2, 3]
  ```

>>>

- **itertools.cycle**: This creates a copy of the iterator and continues to replay the results infinitely:

  ```
  >>> cc = cycle(range(4))
  >>> cc.__next__()
  0
  >>> cc.__next__()
  1
  >>> cc.__next__()
  2
  >>> cc.__next__()
  3
  >>> cc.__next__()
  0
  >>> cc.__next__()
  1
  >>> cc.__next__()
  ```

- **itertools.tee(iterable,number)**: This returns n independent iterables from a single iterable:

  ```
  >>> i,j = tee(range(10),2)
  >>> i
  <itertools._tee object at 0x7ff38e2b2ec8>
  >>> i.__next__()
  ```

```
0
>>> i.__next__()
1
>>> i.__next__()
2
>>> j.__next__()
0
```

- **functools.lru_cache**: This decorator uses memorizing. It saves the results that are mapped to arguments. Hence, it is very useful to speed up functions that take a similar argument, and whose results are not dependent on time or state:

```
In [7]: @lru_cache(maxsize=None)
def fib(n):
    if n<2:
        return n
    return fib(n-1) + fib(n-2)
   ...:

In [8]: %timeit fib(30)
10000000 loops, best of 3: 105 ns per loop

In [9]:
def fib(n):
    if n<2:
        return n
    return fib(n-1) + fib(n-2)
   ...:

In [10]: %timeit fib(30)
1 loops, best of 3: 360 ms per loop
```

- **functools.wraps**: We have just seen how to create decorators, and how to wrap functions. The returned function from decorator retains its name and attributes, such as docstrings, which is not helpful for the users or fellow developers. We can use this decorator to match the returned function to the decorated function. The following snippet shows how it is used:

```
>>> def deco(func):
...     @wraps(func) # this will update wrapper to match func
...     def wrapper(*args, **kwargs):
...         """i am imposter"""
...         print("wrapper")
...         return func(*args, **kwargs)
...     return wrapper
```

```
. . .
>>> @deco
... def realfunc(*args,**kwargs):
...     """i am real function """
...     print("realfunc",args,kwargs)
. . .
>>> realfunc(1,2)
wrapper realfunc
(1, 2) {}
>>> print(realfunc.__name__,
realfunc.__doc__) realfunc i am real function
```

- **Lambda functions**: These functions are simple anonymous functions.
  Lambda functions cannot have statements or annotations. They are very
  useful in creating closures and callbacks in GUI programming:

```
>>> def log(prefix):
...     return lambda x:'%s : %s'%(prefix,x)
. . .
>>> err = log("error")
>>> warn = log("warn")
```

>>>

```
>>> print(err("an error occurred"))
error : an error occurred
>>> print(warn("some thing is not right"))
warn : some thing is not right
```

Sometimes, lambda functions make code easy to understand.

The following is a small program to create the diamond pattern using
the iterations technique and the lambda function:

```
>>> import itertools
>>> af = lambda x:[i for i in itertools.chain(range(1,x+1),range
(x-1,0,-1))]
>>> output = '\n'.join(['%s%s'%(' '*(5-i),' '.join([str(j) for j
in af(i)])) for i in af(5)])
>>> print(output)
        1
      1 2 1
    1 2 3 2 1
  1 2 3 4 3 2 1
1 2 3 4 5 4 3 2 1
  1 2 3 4 3 2 1
    1 2 3 2 1
      1 2 1
        1
```

Summary

In this chapter, we covered how to define functions and pass arguments to them. Then, we discussed decorators in detail; decorators are very popular in frameworks. Toward the end, we collected various utilities that are available in Python, which makes coding a little easier for us.

In the next chapter, we will discuss algorithms and data structures.

4

Data Structures and Algorithms

Data structures are the building blocks to solve programming problems. They provide organization for the data, and algorithms provide the logic to carve the perfect solution. Python provides many efficient built-in data structures that can be used effectively. There are other good data-structure implementations in the standard library as well as third-party libraries. Often, the more pressing question is when to use what, or what data-structure is good for the present problem description. To resolve this, we will cover the following topics:

- Python data structures
- Python library data structures
- Third-party data structures
- Algorithms on scale

Python built-in data structures

Key 1: Understanding Python's in-built data structure.

Before going in on how to use different data structures, we should take a look at the attributes of the object that are important for built-in data structures. For the default sorting to work, the object should have one of the __lt__, and __gt__ methods defined. Otherwise, we can pass a key function to the sorting method to use in getting the intermediate keys that are used to compare it, as shown in the following code:

```
def less_than(self, other):
    return self.data <= other.data
```

```
class MyDs(object):

    def __init__(self, data):
        self.data = data

    def __str__(self,):
        return str(self.data)
    __repr__ = __str__

if __name__ == '__main__':

    ml = [MyDs(i) for i in range(10, 1, -1)]
    try:
        ml.sort()
    except TypeError:
        print("unable to sort by default")

    for att in '__lt__', '__le__', '__gt__', '__ge__':
        setattr(MyDs, att, less_than)
        ml = [MyDs(i) for i in list(range(5, 1, -1)) + list(range(1,
5,))]
        try:
            ml.sort()
            print(ml)
        except TypeError:
            print("cannot sort")
        delattr(MyDs, att)

    ml = [MyDs(i) for i in range(10, 1, -1)]   ·
    print("sorted", sorted(ml, key=lambda x:
x.data)) ml.sort(key=lambda x: x.data)
    print("sort", ml)
```

The output of the preceding code is as follows:

```
[1, 2, 2, 3, 3, 4, 4, 5]
cannot sort
[5, 4, 4, 3, 3, 2, 2, 1]
cannot sort
sorted [2, 3, 4, 5, 6, 7, 8, 9, 10]
sort [2, 3, 4, 5, 6, 7, 8, 9, 10]
```

Whether two objects are equal in value is defined by the output of the __eq__ method. Collections have the same value if they have the same length and the same value of all items, as shown in the following code:

```
def equals(self, other):
    return self.data == other.data


class MyDs(object):

    def __init__(self, data):
        self.data = data

    def __str__(self,):
        return str(self.data)
    __repr__ = __str__

if __name__ == '__main__':
    m1 = MyDs(1)
    m2 = MyDs(2)
    m3 = MyDs(1)
    print(m1 == m2)
    print(m1 == m3)

    setattr(MyDs, '__eq__', equals)
    print(m1 == m2)
    print(m1 == m3)
    delattr(MyDs, '__eq__')

    print("collection")
    l1 = [1, "arun", MyDs(3)]
    l2 = [1, "arun", MyDs(3)]
    print(l1 == l2)
    setattr(MyDs, '__eq__', equals)
    print(l1 == l2)
    l2.append(45)
    print(l1 == l2)
    delattr(MyDs, '__eq__')

    print("immutable collection")
    t1 = (1, "arun", MyDs(3), [1, 2])
    t2 = (1, "arun", MyDs(3), [1, 2])
    print(t1 == t2)
    setattr(MyDs, '__eq__', equals)
```

```
      print(t1 == t2)
      t1[3].append(7)
      print(t1 == t2)
```

The output of the preceding code is as follows:

```
False
False
False
True
collection
False
True
False
immutable collection
False
True
False
```

Hash function maps a larger value set to the smaller hash set. Hence, two different objects can have the same hash, but objects with a different hash must be different. In other words, equal value objects should have the same hash, and objects with a different hash must have different values for hash to be meaningful. When we define __eq__ in a class, we must define a hash function as well. By default, for user class instances, hash uses the ID of the object, as shown in the following code:

```
class MyDs(object):

    def __init__(self, data):
        self.data = data

    def __str__(self,):
        return "%s:%s" % (id(self) % 100000, self.data)

    def __eq__(self, other):
        print("collision")
        return self.data == other.data

    def __hash__(self):
        return hash(self.data)

    __repr__ = __str__
```

```
if __name__ == '__main__':

    dd = {MyDs(i): i for i in (1, 2,
    1)} print(dd)

    print("all collisions")
    setattr(MyDs, '__hash__', lambda x: 1)
    dd = {MyDs(i): i for i in (1, 2,
    1)} print(dd)

    print("all collisions,all values same")
    setattr(MyDs, '__eq__', lambda x, y: True)
    dd = {MyDs(i): i for i in (1, 2,
    1)} print(dd)
```

The output of the preceding code is as follows:

```
collision
{92304:1: 1, 92360:2: 2}
all collisions
collision
collision
{51448:1: 1, 51560:2: 2}
all collisions,all values same
{92304:1: 1}
```

It can be seen that mutable objects do not have hash defined. Although this is not advised, we can, however, do so in our user defined classes:

- **Tuples**: These are immutable lists, slice operations are $O(n)$, retrieval is $O(n)$, and they have small memory requirements. They are normally used to group objects of different types in a single structure, such as C language structures, where the position is fixed for particular types of information, shown as follows:
  ```
  >>> sys.getsizeof(())
  48
  >>> sys.getsizeof(tuple(range(100)))
  848
  ```

  Named tuples that are available from the collections module can be used to access values with object notation, as follows:
  ```
  >>> from collections import namedtuple
  >>> student = namedtuple('student','name,marks')
  >>> s1 = student('arun',133)
  >>> s1.name
  'arun'
  ```

```
>>> s1.marks
133
>>> type(s1)
<class '__main__.student'>
```

- **Lists** : These are mutable data structures that are similar to tuples. They are good to collect objects of similar types. When analyzing their time-complexity, we see that insert, delete, slice, and copy operations require *O(n)*, Retrieval require len *O(1)*, and sort requires *O(nlogn)*. Lists are implemented as dynamic arrays. It must resize to double of its previous on increase in size greater than current capacity. Insert and delete at the front of the list takes more time as it must move all references to other elements one by one:

```
>>> sys.getsizeof([])
64
>>> sys.getsizeof(list(range(100)))
1008
```

- **Dictionary**: These are mutable mappings. A key can be any hashable object. Getting a value for key, setting a value for a key, and deleting a key are all *O(1)*, and copying is *O(n)*:

```
>>> d = dict()
>>> getsizeof(d)
288
>>> getsizeof({i:None for i in range(100)})
6240
```

- **Sets**: These can be thought as of groups of items where hash is used to retrieve items. Sets have methods to check union, and intersection, which is useful rather than checking the same with lists. Let's take an example of groups of animals, as follows:

```
>>> air = ("sparrow", "crow")
>>> land = ("sparrow","lizard","frog")
>>> water = ("frog","fish")
>>> # find animal able to live on land and water
...
>>> [animal for animal in water if animal in
land] ['frog']

>>> air = set(air)
>>> land = set(land)
>>> water = set(water)
>>> land | water #animal living either land or water
{'frog', 'fish', 'sparrow', 'lizard'}
```

```
>>> land & water #animal living both land and
water {'frog'}
>>> land ^ water #animal living on only one land or water
{'fish', 'sparrow', 'lizard'}
```

Their implementation and time-complexity is very similar to dictionary, shown as follows:

```
>>> s = set()
>>> sys.getsizeof(s)
224
>>> s = set(range(100))
>>> sys.getsizeof(s)
8416
```

Python library data structures

Key 2: Using Python's standard library data structures.

- **collections.deque**: The collections module have a `deque` implementation. Deque is useful for the scenarios where item insertion and deletion occurs at both ends of structure as it has efficient inserts at the start of structure as well. Time-complexity is similar to copy *O(n)*, insert—*O(1)*, and delete—*O(n)*. The following graph shows an insert at 0 position operation comparison between list and deque:

```
>>> d = deque()
>>> getsizeof(d)
632
>>> d = deque(range(100))
>>> getsizeof(d)
1160
```

The following image is the graphical representation of the preceding code:

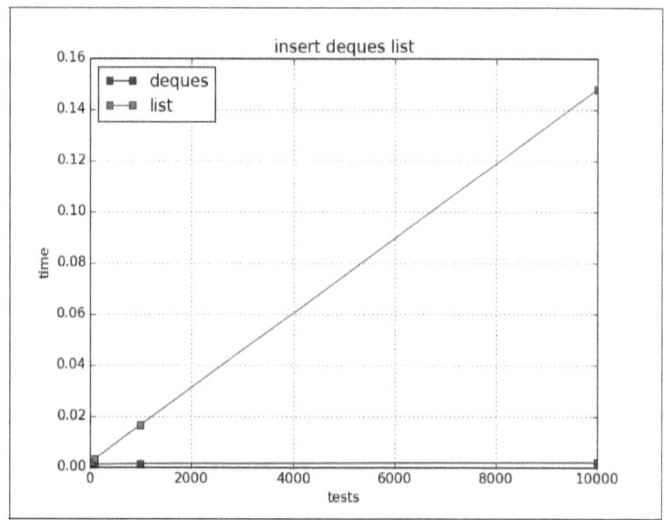

- **PriorityQueue**: A standard library queue module has implementations for multiproducer, and multiconsumer queues. We can simplify and reuse its PriorityQueue for simpler cases using the heapq module, as follows:

```
from heapq import heappush, heappop
from itertools import count

class PriorityQueue(object):
    def __init__(self,):
        self.queue = []
        self.counter = count()

    def __len__(self):
        return len(self.queue)

    def pop(self,):
        item = heappop(self.queue)
        print(item)
```

```
            return item[2],item[0]

    def push(self,item,priority): cnt =
        next(self.counter) heappush(self.queue,
        (priority, cnt, item))
```

Other than these, queue modules have `threadsafe`, `LifoQueue`, `PriorityQueue`, `queue`, `deque` implementations. Also, lists can be used as stacks or queues. Collections also have `orderedDict`, which remembers the sequence of elements.

Third party data structures

Key 3: Using third-party data structures.

Python has a good bunch of data structures in the core language/library. But sometimes, an application has very specific requirements. We can always use third-party data-structure packages. Most of such modules are Python wrapper over C, C++ implementations:

- The `blist` module provides a drop-in replacement for list, `sortedList`, and `sortedset`. It is discussed in greater detail in later chapters.
- The `bintrees` module provides binary, AVL tree, and Red-Black trees.
- The `banyan` module provides Red-Black trees, splay tree, and sorted lists.
- The `Sortedcontainers` module provides `SortedList`, `SortedDict`, and `SortedSet`. So, one can get almost every data structure for Python easily. More stress should be given on why one data structure is better than another for a use case.

Arrays/List

For numeric calculations involving math, NumPy arrays should be considered. They are fast, memory-efficient, and provide many vector and matrix operations.

Binary tree

Trees have better worst-case insertion/removal, $O(log(n))$, min/max, and look-ups than dictionaries. There are several implementations that are available.

One module is `bintrees`, which have C implementations that are available for Red-Black trees, AVL tree, and Binary trees. For example, in Red-Black trees, it is easy to find max, and min, ranges as shown in the following example:

```
tr = bintrees.FastRBTree()
tr.insert("a",40)
tr.insert("b",5)
tr.insert("a",9)
print(list(tr.keys()),list(tr.items()))
print(tr.min_item())
print(tr.pop_max())
print(tr.pop_max())
tr = bintrees.FastRBTree([(i,i+1) for i in range(10)])
print(tr[5:9])
```

The output of the preceding code is as follows:

```
['a', 'b'] [('a', 9), ('b', 5)]
('a', 9)
('b', 5)
('a', 9)
FastRBTree({5: 6, 6: 7, 7: 8, 8: 9})
```

Sorted containers

These are pure-Python modules having `SortedList`, `SortedSet`, and `SortedDict` Data structures, which can keep the keys/items sorted. The `SortedContainers` module claims to have speed comparable to C extensions modules, shown as follows:

```
import sortedcontainers as sc
import sys
l = sc.SortedList()
l.update([0,4,2,1,4,2])
print(l)
print(l.index(2),l.index(4))
l.add(6)
print(l[-1])
l = sc.SortedList(range(10))
print(l)
print(list(l.irange(2,6)))

seta = sc.SortedSet(range(1,4))
setb = sc.SortedSet(range(3,7))
print(seta - setb)
```

```
print (seta | setb )
print (seta & setb)
print ([i for i in seta])
```

The output of the preceding code is as follows:

```
SortedList([0, 1, 2, 2, 4, 4], load=1000)
2 4

SortedList([0, 1, 2, 3, 4, 5, 6, 7, 8, 9],
load=1000) [2, 3, 4, 5, 6]
SortedSet([1, 2], key=None, load=1000)
SortedSet([1, 2, 3, 4, 5, 6], key=None,
load=1000) SortedSet([3], key=None, load=1000)
[1, 2, 3]
```

Trie

This is an ordered-tree data-structure, where the position in the tree defines the key. The keys are normally strings. In comparison to dictionaries, it has faster worst-case data retrieval $O(m)$. Hash functions are not needed. If we are using strings only to be stored in the keys, it can take a lot less space then dictionary.

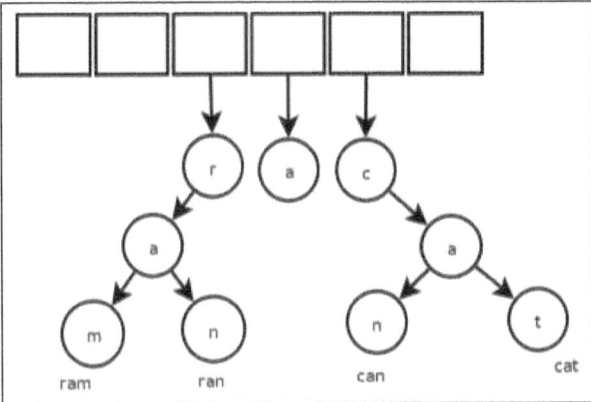

In Python, we have the `marisa-trie` package that provides this functionality as static Data structures. It is a Cython wrapper over the C++ library. We can associate values with the keys as well. It also provides memory mapped I/O, which is useful to decrease memory usage on cost of speed. The `datrie` is another package that provides read-write tries, The following are some basic usage of these libraries:

```
>>> import string
>>> import marisa_trie as mtr
>>> import datrie as dtr
>>>
>>>
>>> # simple read-only keys
... tr = mtr.Trie([u'192.168.124.1',u'192.168.124.2',u'10.10.1.1'
,u'10.10.1.2'])
>>> #get all keys
... print(tr.keys())
['10.10.1.1', '10.10.1.2', '192.168.124.1', '192.168.124.2']
>>> #check if key exists
... print(tr.has_keys_with_prefix('192'))
True
>>> # get id of key
... print(tr.get('192.168.124.1'))
2
>>> # get all items
... print(tr.items())
[('10.10.1.1', 0), ('10.10.1.2', 1), ('192.168.124.1',
2), ('192.168.124.2', 3)]
>>>

>>> # storing data along with keys
... btr = mtr.BytesTrie([('192.168.124.1',b'redmine.meeku.com'),

... ('192.168.124.2',b'jenkins.meeku.com'), ...
('10.5.5.1',b'gerrit.chiku.com'), ...
('10.5.5.2',b'gitlab.chiku.com'), ... ])

>>> print(list(btr.items()))
[('10.5.5.1',          b'gerrit.chiku.com'),          ('10.5.5.2',
b'gitlab.chiku.com'),     ('192.168.124.1',     b'redmine.meeku.com'),
('192.168.124.2', b'jenkins. meeku.com')]
>>> print(btr.get('10.5.5.1'))
[b'gerrit.chiku.com']
>>>

>>> with open("/tmp/marisa","w") as f:
...btr.write(f)
...
>>>
```

```
... # using memory mapped io to decrease memory
usage ... dbtr = mtr.BytesTrie().mmap("/tmp/marisa")
>>> print(dbtr.get("192.168.124.1"))
[b'redmine.meeku.com']
```

>>>
>>>

```
>>> trie = dtr.Trie('0123456789.') #define allowed character range
>>> trie['192.168.124.1']= 'redmine.meeku.com'
>>> trie['192.168.124.2'] = 'jenkins.meeku.com'
>>> trie['10.5.5.1'] = 'gerrit.chiku.com'
>>> trie['10.5.5.2'] = 'gitlab.chiku.com'
>>> print(trie.prefixes('192.168.245'))
[]
>>> print(trie.values())
['gerrit.chiku.com', 'gitlab.chiku.com',
'redmine.meeku.com', 'jenkins.meeku.com']
>>> print(trie.suffixes())
['10.5.5.1', '10.5.5.2', '192.168.124.1', '192.168.124.2']
>>>
>>> trie.save("/tmp/1.datrie")
>>> ntr = dtr.Trie.load('/tmp/1.datrie')
>>> print(ntr.values())
['gerrit.chiku.com', 'gitlab.chiku.com',
'redmine.meeku.com', 'jenkins.meeku.com']
>>> print(ntr.suffixes())
['10.5.5.1', '10.5.5.2', '192.168.124.1', '192.168.124.2']
```

Algorithms on scale

Key 4: Thinking out-of-the-box for the algorithms.

An algorithm is how we solve a problem. The most common issue of not being able to solve the problem is to not being able to define it properly. Normally, we look to apply an algorithm only at a small level, such as in a small functionality, or sorting in a function. We, however, do not think about algorithms when the scale increases, mostly the stress is on how fast it is. Let's take a simple requirement of sorting a file and sending it to a user. If the file is, let's say 10-20 KB or so, it will be best to simply use the Python sorted function to sort the entries. In the following code, the file is of format where columns are ID, name, due, and due-date. We want to sort it based on dues, as follows:

```
10000000022,shy-flower-ac5,-473,16/03/25
10000000096,red-water-e85,-424,16/02/12
10000000097,dry-star-c85,-417,16/07/19
```

```
10000000070,damp-night-c76,-364,16/03/12
10000000032,muddy-shadow-aad,-362,16/08/05

def dosort(filename,result):
    with open(filename) as ifile:
        with open(result,"w") as ofile:
            for line in sorted(
                map(lambda x:x.strip(), ifile.readlines()
                    ),key=lambda x:int(x.split(',')[2])
                ):
                ofile.write(line)
                ofile.write('\n')
```

This works great, but as the file increases in size, the memory requirement increases. We cannot load all contents in the memory at the same time. Hence, we can use external merge-sort to divide the file into small parts, sort them, and then merge the sorted results together. In the following code, we used `heapq.merge` to merge iterators:

```
import tempfile
import heapq

def slowread(f, nbytes):
    while True:
        ilines = f.readlines(nbytes)
        if not ilines:
            break
        for line in ilines:
            yield int(line.split(',')[2]),line

def dosort(filename, result):
    partition = 5000
    with open(filename,"r") as ifile:
        with open(result,"w") as ofile:
            tempfiles = []
            while True:
                ilines = ifile.readlines(partition)
                if len(ilines) == 0 :
                    break
                tfile = tempfile.TemporaryFile(mode="w+")
                tfile.writelines(
                    sorted(
                        ilines,
                        key=lambda x:int(x.split(',')[2])
                        ))
```

```
            tfile.seek(0)
            tempfiles.append(tfile)
        lentempfiles = len(tempfiles)
        read_generators = [slowread(tfile,
partition// (lentempfiles+1)) for tfile in tempfiles]
        res = []
        for line in heapq.merge(*read_generators):
            res.append(line[1])
            if len(res) > 100:
                ofile.writelines(res)
                res.clear()
        if res:
            ofile.writelines(res)
        ofile.close()
```

Once we use up memory of a single computer, or have files distributed over multiple computers in a network, the file-based algorithm will not work. We will need to sort incoming streams from upstream servers, and send the sorted stream to the downstream. If we look at the following code carefully, we will see that we have not changed the underlying mechanism. We are still using `heapq.merge` to merge elements, but now, we are getting elements from the network instead. The following client code is simple, it just starts sending sorted lines by lines on receive of the next command from a downstream server:

```
import socket
import sys
from sort2 import dosort2

HOST = '127.0.0.1'
PORT =  9002
NCLIENTS = 2

class Client(object):

    def __init__(self,HOST,PORT,filename):
        self.skt = socket.socket(socket.AF_INET,socket.SOCK_STREAM)
        self.skt.connect((HOST,PORT))
        self.filename = filename
        self.skt.setblocking(True)

    def run(self):
        for line in dosort2(self.filename):
            print("for",line)
            data = self.skt.recv(1024)
```

```
              print("data cmd",data)
              if data == b'next\r\n':
                  data = None
                  self.skt.send(line[1].encode())
              else:
                  print("got from server",data)
          print("closing socket")
          self.skt.close()

c = Client(HOST,PORT,sys.argv[1])
c.run()
```

In the server code, the `ClientConn` **class abstracts away network operations and provides an iterator interface to** `heapq.merge`. We can greatly enhance the code using buffering. Here, the `get_next` **method gets new line from the client. Simple abstraction solved a great problem:**

```
import socket
import heapq
from collections import deque

HOST = '127.0.0.1'
PORT = 9002
NCLIENTS = 2

class Empty(Exception):
    pass

class ClientConn(object):
    def __init__(self, conn, addr):
        self.conn = conn
        self.addr = addr
        self.buffer = deque()
        self.finished = False
        self.get_next()

    def __str__(self, ):
        return '%s' % (str(self.addr))

    def get_next(self):
        print("getting next", self.addr)
        self.conn.send(b"next\r\n")
        try:
            ndata = self.conn.recv(1024)
```

```
            except Exception as e:
                print(e)
                self.finished = True
                ndata = None
            if ndata:
                ndata = ndata.decode()
                print("got from client", ndata)
                self.push((int(ndata.split(',')[2]), ndata))
            else:
                self.finished = True

    def pop(self):
        if self.finished:
            raise Empty()
        else:
            elem = self.buffer.popleft()
            self.get_next()
            return elem

    def push(self, value):
        self.buffer.append(value)

    def __iter__(self, ):
        return self

    def __next__(self, ):
        try:
            return self.pop()
        except Empty:
            print("iter empty")
            raise StopIteration


class Server(object):
    def __init__(self, HOST, PORT, NCLIENTS):
        self.nclients = NCLIENTS
        self.skt = socket.socket(socket.AF_INET, socket.SOCK_STREAM)
        self.skt.setblocking(True)
        self.skt.bind((HOST, PORT))
        self.skt.listen(1)

    def run(self):
```

```
        self.conns = []  # list of all clients connected

        while len(self.conns) < self.nclients: # accept client till
we have all
            conn, addr = self.skt.accept()
            cli = ClientConn(conn, addr)
            self.conns.append(cli)
            print('Connected by', cli)

        with open("result", "w") as ofile:
            for line in heapq.merge(*self.conns):
                print("output", line)
                ofile.write(line[1])

s = Server(HOST, PORT, NCLIENTS)
s.run()
```

Summary

In this chapter, we learned about the data structures that are available in the Python standard library and some third-party libraries, which are extremely useful for everyday programming. Knowledge of data-structure usage is very much important in choosing right tool for the job. Choosing of an algorithm is highly application-specific, and we should always try to find out a solution that is simpler to read.

In the next chapter, we will cover design patterns that provide great help in writing elegant solutions to the problems.

5

Elegance with
Design Patterns

In this chapter, we are going to learn some design patterns that will help us in writing better software, which is reusable and tidy. But, the biggest help is that they let developers think on an architectural level. They are solutions to recurring problems. While learning them is very helpful for compiled languages such as C and C++ because they are actually solutions to problems, in Python, developers often "just write code" without needing any design pattern due to the dynamism in the language and conciseness of code. This is largely true for developers whose first language is Python. My advice is to learn design patterns to be able to process information and design at an architectural level rather than function and classes.

In this chapter, we will cover the following topics:

- Observer pattern
- Strategy pattern
- Singleton pattern
- Template pattern
- Adaptor pattern
- Facade pattern
- Flyweight pattern
- Command pattern
- Abstract factory
- Registry pattern
- State pattern

Observer pattern

Key 1: Spreading information to all listeners.

This is the basic pattern in which an object tells other objects about something interesting. It is very useful in GUI applications, pub/sub applications, and those applications where we need to notify a lot of loosely-coupled application components about a change occurring at one source node. In the following code, Subject is the object to which other objects register themselves for events via register_observer. The observer objects are the listening objects. The observers start observing the function that registers the observers object to Subject object. Whenever there is an event to Subject it cascades the event to all observers:

```python
import weakref


class Subject(object):
    """Provider of notifications to other
    objects """

    def __init__(self, name):
        self.name = name
        self._observers = weakref.WeakSet()

    def register_observer(self, observer):
        """attach the observing object for this subject
        """
        self._observers.add(observer)
        print("observer {0} now listening on
            {1}".format( observer.name, self.name))

    def notify_observers(self, msg):
        """transmit event to all interested
        observers """
        print("subject notifying observers about {}".format(msg,))
        for observer in self._observers:
            observer.notify(self, msg)


class Observer(object):

    def __init__(self, name):
```

```
        self.name = name

    def start_observing(self, subject):
        """register for getting event for a subject
        """
        subject.register_observer(self)

    def notify(self, subject, msg):
        """notify all observers
        """
        print("{0} got msg from {1} that {2}".format(
            self.name, subject.name, msg))

class_homework = Subject("class homework")
student1 = Observer("student 1")
student2 = Observer("student 2")

student1.start_observing(class_homework)
student2.start_observing(class_homework)

class_homework.notify_observers("result is out")

del student2

class_homework.notify_observers("20/20 passed this sem")
```

The output for the preceding code is as follows:

```
(tag)[ ch5 ] $ python codes/B04885_05_code_01.py
observer student 1 now listening on class homework
observer student 2 now listening on class homework
subject notifying observers about result is out
student 1 got msg from class homework that result is out
student 2 got msg from class homework that result is out
subject notifying observers about 20/20 passed this sem
student 1 got msg from class homework that 20/20 passed this sem
```

Strategy pattern

Key 2: Changing the behavior of an algorithm.

Sometimes, the same piece of code must have different behavior for different invocation by different clients. For example, time-conversion for all countries must handle daylight-savings time in some countries and change their strategy in these cases. The main use is to switch the implementation. In this pattern, algorithm's behavior is selected on runtime. As Python is a dynamic language, it is trivial to assign functions to variables and change them on runtime. Similar to the following code segment, there are two implementations to calculate tax, namely, `tax_simple`, and `tax_actual`. For the following code snippet, `tax_cal` references clients that are used. The implementation can be changed by changing reference to the implementing function:

```
TAX_PERCENT = .12

def tax_simple(billamount):
    return billamount * TAX_PERCENT

def tax_actual(billamount):
    if billamount < 500:
        return billamount * (TAX_PERCENT//2)
    else:
        return billamount * TAX_PERCENT

tax_cal = tax_simple
print(tax_cal(400),tax_cal(700))

tax_cal = tax_actual
print(tax_cal(400),tax_cal(700))
```

The output of the preceding code snippet is as follows:

```
48.0 84.0
0.0 84.0
```

But the issue with the preceding implementation is that at one time all clients will
see the same strategy for tax calculation. We can improve this using a class that
selects the implementation based on request parameters. In the following example,
in the `TaxCalculator` class's instance, the strategy is determined for each call to it
on runtime. If the request is for India IN, Tax is calculated as per the Indian
standard, and if request is for US, it is calculated as per the US standard:

```
TAX_PERCENT = .12

class TaxIN(object):
    def __init__(self,):
        self.country_code = "IN"

    def __call__(self, billamount):
        return billamount * TAX_PERCENT

class TaxUS(object):
    def __init__(self,):
        self.country_code = "US"

    def __call__(self,billamount):
        if billamount < 500:
            return billamount * (TAX_PERCENT//2)
        else:
            return billamount * TAX_PERCENT

class TaxCalculator(object):

    def __init__(self):
        self._impls = [TaxIN(),TaxUS()]

    def __call__(self, country, billamount):
    """select the strategy based on country
    parameter """
        for impl in self._impls:
            if impl.country_code == country:
                return impl(billamount)
            else:
                return None

tax_cal = TaxCalculator()
print(tax_cal("IN", 400), tax_cal("IN", 700))
print(tax_cal("US", 400), tax_cal("US", 700))
```

The output of the preceding code is as follows:

```
48.0 84.0
0.0 84.0
```

Singleton pattern

Key 3: Providing the same view to all.

The singleton pattern maintains the same state for all instances of a class. When we change an attribute at one place in a program, it is reflected in all references to this instance. As modules are globally shared, we can use them as singleton methods, and the variables defined in them are the same everywhere. But, there are similar issues in that as the module is reloaded, there may be more singleton classes that are needed. We can also create a singleton pattern using metaclasses in the following manner. The six is a third-party library to help in writing the same code that is runnable on Python 2 and Python 3.

In the following code, Singleton metaclass has a registry dictionary where the instance corresponding to each new class is stored. When any class asks for a new instance, this class is searched for in the registry, and if found, the old instance is passed. Otherwise, a new instance is created, stored in registry, and returned. This can be seen in the following code:

```
from six import with_metaclass


class Singleton(type):
    _registry = {}

    def __call__(cls, *args, **kwargs):
        print(cls, args, kwargs)
        if cls not in Singleton._registry:
            Singleton._registry[cls] = type.__call__(cls, *args,
**kwargs)
        return Singleton._registry[cls]


class Me(with_metaclass(Singleton, object)):

    def __init__(self, data):
        print("init ran", data)
```

```
        self.data = data

m = Me(2)
n = Me(3)
print(m.data, n.data)
```

The following is the output of the preceding code:

```
<class '__main__.Me'> (2,) {}
init ran 2
<class '__main__.Me'> (3,) {}
2 2
```

Template pattern

Key 4: Refining algorithm to use case.

In this pattern, we define the skeleton of an algorithm in a method called the
`template` method, which defers some of its steps to subclasses. How we do this is
as follows, we analyze the procedure, and break it down to logical steps, which are
different for different use cases. Now, we may or may not implement the default
implementation of these steps in the main class. The subclasses of the main class will
implement the steps that are not implemented in the main class, and they may skip
some generic steps implementation. In the following example, `AlooDish` is base
class with the `cook` template method. It applies to normal Aloo fried dishes, which
have a common cooking procedure. Each recipe is a bit different in ingredients, time
to cook, and so on. Two variants, `AlooMatar`, and `AlooPyaz`, define some set of steps
differently than others:

```
import six

class AlooDish(object):

    def get_ingredients(self,):
        self.ingredients = {}

    def prepare_vegetables(self,):
        for item in six.iteritems(self.ingredients):
            print("take {0} {1} and cut into smaller pieces".
format(item[0],item[1]))
        print("cut all vegetables in small pieces")

    def fry(self,):
```

```
            print("fry for 5 minutes")

    def serve(self,):
        print("Dish is ready to be served")

    def cook(self,):
        self.get_ingredients()
        self.prepare_vegetables()
        self.fry()
        self.serve()

class AlooMatar(AlooDish):

    def get_ingredients(self,):
        self.ingredients = {'aloo':"1 Kg",'matar':"1/2 kg"}

    def fry(self,):
        print("wait 10 min")

class AlooPyaz(AlooDish):

    def get_ingredients(self):
        self.ingredients = {'aloo':"1 Kg",'pyaz':"1/2 kg"}

aloomatar = AlooMatar()
aloopyaz = AlooPyaz()
print("******************  aloomatar cook")
aloomatar.cook()
print("****************** aloopyaz cook")
aloopyaz.cook()
```

The following is the output of the preceding example code:

```
******************  aloomatar cook
take matar 1/2 kg and cut into smaller
pieces take aloo 1 Kg and cut into smaller
pieces cut all vegetables in small pieces
wait 10 min
Dish is ready to be served
****************** aloopyaz cook
take pyaz 1/2 kg and cut into smaller pieces
take aloo 1 Kg and cut into smaller pieces
cut all vegetables in small pieces fry for 5
minutes
Dish is ready to be served
```

Adaptor pattern

Key 5: Bridging class interfaces.

This pattern is used to adapt a given class to a new interface. It solves the problem for an interface mismatch. To demonstrate this, let's assume that we have an API function that creates a competition to run different animals. Animals should have a `running_speed` function, which tells their speed to compare them. `Cat` is one such class. Now, if we have a `Fish` class in a different library, which also wants to participate in the function, it must be able to know its `running_speed` function. As changing the implementation of `Fish` is not good option, we can create an `adaptor` class that can adapt the `Fish` class to run by providing the necessary bridge:

```python
def running_competition(*list_of_animals):
    if len(list_of_animals)<1:
        print("No one Running")
        return
    fastest_animal = list_of_animals[0]
    maxspeed = fastest_animal.running_speed()
    for animal in list_of_animals[1:]:
        runspeed =  animal.running_speed()
        if runspeed > maxspeed:
            fastest_animal = animal
            maxspeed = runspeed
    print("winner is {0} with {1} Km/h".format(fastest_animal.
name,maxspeed))


class Cat(object):

    def __init__(self, name, legs):
        self.name = name
        self.legs = legs

    def running_speed(self,):
        if self.legs>4 :
            return 20
        else:
            return 40


running_competition(Cat('cat_a',4),Cat('cat_b',3))

class Fish(object):
```

```
    def __init__(self, name, age):
        self.name = name
        self.age = age

    def swim_speed(self):
        if self.age < 2:
            return 40
        else:
            return 60

# to let our fish to participate in tournament it should have similar
interface as
# cat, we can also do this by using an adaptor class RunningFish

class RunningFish(object):
    def __init__(self, fish):
        self.legs = 4 # dummy
        self.fish = fish

    def running_speed(self):
        return self.fish.swim_speed()

    def __getattr__(self, attr):
        return getattr(self.fish, attr)

running_competition(Cat('cat_a', 4),
                    Cat('cat_b', 3),
                    RunningFish(Fish('nemo', 3)),
                    RunningFish(Fish('dollar', 1)))
```

The output of the preceding code is follows:

```
winner is cat_a with 40 Km/h
winner is nemo with 60 Km/h
```

Facade pattern

Key 6: Hiding system complexity for a simpler interface.

In this pattern, a main class called facade exports a simpler interface to client classes and encapsulates the complexity of interaction with many other classes of the system. It is like a gateway to a complex set of functionality, such as in the following example, the WalkingDrone class hides the complexity of synchronization of the Leg classes and provides a simpler interface to client classes:

```python
class Leg(object):
    def __init__(self,name):
        self.name = name

    def forward(self):
        print("{0},".format(self.name), end="")


class WalkingDrone(object):

    def __init__(self, name):
        self.name = name
        self.frontrightleg = Leg('Front Right Leg')
        self.frontleftleg = Leg('Front Left Leg')
        self.backrightleg = Leg('Back Right Leg')
        self.backleftleg = Leg('Back Left Leg')

    def walk(self):
        print("\nmoving ",end="")
        self.frontrightleg.forward()
        self.backleftleg.forward()
        print("\nmoving ",end="")
        self.frontleftleg.forward()
        self.backrightleg.forward()

    def run(self):
        print("\nmoving ",end="")
        self.frontrightleg.forward()
        self.frontleftleg.forward()
        print("\nmoving ",end="")
        self.backrightleg.forward()
        self.backleftleg.forward()

wd = WalkingDrone("RoboDrone" )
```

```
print("\nwalking")
wd.walk()
print("\nrunning")
wd.run()
```

This code will give us the following output:

```
walking

moving Front Right Leg,Back Left Leg,
moving Front Left Leg,Back Right Leg,
running

moving Front Right Leg,Front Left Leg,
moving Back Right Leg,Back Left Leg,Summary
```

Flyweight pattern

Key 7: Consuming less memory with shared objects.

A flyweight design pattern is useful to save memory. When we have lots of object count, we store references to previous similar objects and provide them instead of creating new objects. In the following example, we have a Link class used by the browser, which stores the link data.

The browser uses this data, and there may be a lot of data that is associated with pictures referenced by the link, such as image content, size, and so on, and images can be reused over the page. Hence, the nodes using it only store a flyweight BrowserImage object to decrease the memory footprint. When the link class tries to create a new BrowserImage instance, the BrowserImage class checks whether it has an instance in its _resources mapping for the resource path. If it does, it will just pass the old instance:

```
import weakref


class Link(object):

    def __init__(self, ref, text, image_path=None):
        self.ref = ref
        if image_path:
            self.image = BrowserImage(image_path)
        else:
            self.image = None
```

```
            self.text = text

    def __str__(self):
        if not self.image:
            return "<Link (%s)>" % self.text
        else:
            return "<Link (%s,%s)>" % (self.text, str(self.image))


class BrowserImage(object):
    _resources = weakref.WeakValueDictionary()

    def __new__(cls, location):
        image = BrowserImage._resources.get(location, None)
        if not image:
            image = object.__new__(cls)
            BrowserImage._resources[location] = image
            image.__init(location)
        return image

    def __init(self, location):
        self.location = location
        # self.content = load picture into memory

    def __str__(self,):
        return "<BrowserImage(%s)>" % self.location

icon = Link("www.pythonunlocked.com",
            "python unlocked book",
            "http://pythonunlocked.com/media/logo.png")
footer_icon = Link("www.pythonunlocked.com/#bottom",
                   "unlocked series python book",
                   "http://pythonunlocked.com/media/logo.png")
twitter_top_header_icon = Link("www.twitter.com/pythonunlocked",
                               "python unlocked twitter link",
                               "http://pythonunlocked.com/media/logo.
png")

print(icon,)
print(footer_icon,)
print(twitter_top_header_icon,)
```

The output of the preceding code is follows:

```
<Link (python unlocked
book,<BrowserImage(http://pythonunlocked.com/ media/logo.png)>)>
<Link (unlocked series python book,<BrowserImage(http://
pythonunlocked.com/media/logo.png)>)>
<Link (python unlocked twitter link,<BrowserImage(http://
pythonunlocked.com/media/logo.png)>)>
```

Command pattern

Key 8: Easy-execution management for commands.

In this pattern, we encapsulate information that is needed to execute a command in
an object so that command itself can have further capabilities, such as undo, cancel,
and metadata that are needed at a later point of time. For example, let's create a
simple `Chef` in a restaurant, users can issue orders (commands), commands here
have metadata that are needed to cancel them. This is similar to a notepad app
where each user action is recorded with an undo method. This makes coupling
loose between caller and the invoker, shown as follows:

```python
import time
import threading

class Chef(threading.Thread):

    def __init__(self,name):
        self.q = []
        self.doneq = []
        self.do_orders = True
        threading.Thread.__init__(self,)
        self.name = name
        self.start()

    def makeorder(self, order):
        print("%s Preparing Menu :"%self.name )
        for item in order.items:
            print("cooking ",item)
            time.sleep(1)
        order.completed = True
        self.doneq.append(order)


    def run(self,):
```

```
        while self.do_orders:
            if len(self.q) > 0:
                order = self.q.pop(0)
                self.makeorder(order)
                time.sleep(1)

    def work_on_order(self,order):
        self.q.append(order)

    def cancel(self, order):
        if order in self.q:
            if order.completed == True:
                print("cannot cancel, order completed")
                return
            else:
                index = self.q.index(order)
                del self.q[index]
                print(" order canceled %s"%str(order))
                return
        if order in self.doneq:
            print("order completed, cannot be canceled")
            return
        print("Order not given to me")


class Check(object):

    def execute(self,):
        raise NotImplementedError()

    def cancel(self,):
        raise NotImplementedError()

class MenuOrder(Check):

    def __init__(self,*items):
        self.items = items
        self.completed = False

    def execute(self,chef):
        self.chef = chef
        chef.work_on_order(self)

    def cancel(self,):
```

```
        if self.chef.cancel(self):
            print("order cancelled")

    def __str__(self,):
        return ''.join(self.items)

c = Chef("Arun")
order1 = MenuOrder("Omellette", "Dosa", "Idli")
order2 = MenuOrder("Mohito", "Pizza")
order3 = MenuOrder("Rajma", )
order1.execute(c)
order2.execute(c)
order3.execute(c)

time.sleep(1)
order3.cancel()
time.sleep(9)
c.do_orders = False
c.join()
```

The output of the preceding code is as follows:

```
Arun Preparing Menu :
cooking  Omellette
 order canceled Rajma
cooking  Dosa
cooking  Idli
Arun Preparing Menu :
cooking  Mohito
cooking  Pizza
```

Abstract factory

This design pattern creates an interface to create a family of interrelated objects without specifying their concrete class. It is similar to a superfactory. Its advantage is that we can add further variants, and clients will not have to worry further about the interface or actual classes for the new variants. It is helpful in supporting various platforms, windowing systems, data types, and so on. In the following example, the `Animal` class is the interface that the client will know about for any animal instance. `AnimalFactory` is the abstract factory that `DogFactory` and `CatFactory` implement. Now, on the runtime by user input, or configuration file, or runtime environment check, we can decide whether we will have all `Dog` or `Cat` instances. It is very convenient to add a new class implementation, as follows:

```
import os
import abc
```

```
import six


class Animal(six.with_metaclass(abc.ABCMeta, object)):
    """ clients only need to know this interface for
    animals""" @abc.abstractmethod
    def sound(self, ):
        pass

class AnimalFactory(six.with_metaclass(abc.ABCMeta, object)):
    """clients only need to know this interface for
creating animals"""
    @abc.abstractmethod
    def create_animal(self,name):
        pass

class Dog(Animal):
    def __init__(self, name):
        self.name = name

    def sound(self, ):
        print("bark bark")

class DogFactory(AnimalFactory):
    def create_animal(self,name):
        return Dog(name)

class Cat(Animal):
    def __init__(self, name):
        self.name = name
    def sound(self, ):
        print("meow meow")

class CatFactory(AnimalFactory):
    def create_animal(self,name):
        return Cat(name)

class Animals(object):
    def __init__(self,factory):
        self.factory = factory

    def create_animal(self, name):
```

```
                    return self.factory.create_animal(name)


if __name__ == '__main__':
    atype = input("what animal (cat/dog) ?").lower()
    if atype == 'cat':
        animals = Animals(CatFactory())
    elif atype == 'dog':
        animals = Animals(DogFactory())
    a = animals.create_animal('bulli')
    a.sound()
```

The preceding code will give us the following output:

```
1st run:

what animal (cat/dog) ?dog
bark bark

2nd run:
what animal (cat/dog) ?cat
meow meow
```

Registry pattern

Key 9: Adding functionality from anywhere in code to class.

This is one of my favorite patterns and comes to help a lot. In this pattern, we register classes to a registry, which tracks the naming to functionality. Hence, we can add functionality to the main class from anywhere in the code. In the following code, Convertor tracks all convertors from dictionary to Python objects. We can easily add further functionalities to the system using the convertor.register decorator from anywhere in the code, as follows:

```
class ConvertError(Exception):

    """Error raised on errors on conversion"""
    pass


class Convertor(object):

    def __init__(self,):
```

```
        """create registry for storing method mapping """
        self.__registry = {}

    def to_object(self, data_dict):
        """convert to python object based on type of dictionary"""
        dtype = data_dict.get('type', None)
        if not dtype:
            raise ConvertError("cannot create object, type not
defined")
        elif dtype not in self.__registry:
            raise ConvertError("cannot convert type not registered")
        else:
            convertor = self.__registry[dtype]
            return convertor.to_python(data_dict['data'])

    def register(self, convertor):
        iconvertor = convertor()
        self.__registry[iconvertor.dtype] = iconvertor


convertor = Convertor()


class Person():

    """ a class in application """

    def __init__(self, name, age):
        self.name = name
        self.age = age

    def __str__(self,):
        return "<Person (%s, %s)>" % (self.name, self.age)


@convertor.register
class PersonConvertor(object):

    def __init__(self,):
        self.dtype = 'person'

    def to_python(self, data):
```

```
              # not checking for errors in dictionary to instance
              creation p = Person(data['name'], data['age'])
              return p

      print(convertor.to_object(
          {'type': 'person', 'data': {'name': 'arun', 'age': 12}}))
```

The following is the output for the preceding code:

```
      <Person (arun, 12)>
```

State pattern

Key 10: Changing execution based on state.

State machines are very useful for an algorithm whose vector-flow of control depends on the state of the application. Similar to when parsing a log output with sections, you may want to change the parser logic on every next section. It is also very useful to write code for network servers/clients who enable certain commands in a certain scope:

```
      def outputparser(loglines):
          state = 'header'
          program, end_time, send_failure=
          None, None, False for line in loglines:
              if state == 'header':
                  program = line.split(',')[0]
                  state = 'body'
              elif state == 'body':
                  if 'send_failure' in line:
                      send_failure = True
                  if '======' in line:
                      state = 'footer'
              elif state == 'footer':
                  end_time = line.split(',')[0]
          return program, end_time, send_failure

      print(outputparser(['sampleapp,only a sampleapp',
                  'logline1  sadfsfdf',
                  'logline2 send_failure',
                  '=====================',
                  '30th Jul 2016,END']))
```

This will give us the following output:

```
('sampleapp', '30th Jul 2016', True)
```

Summary

In this chapter, we saw various design patterns that can help us better organize the code, and in some cases, increase performance. The good thing about patterns is they let you think beyond classes, and they provide strategy for architecture of your application. As closing advice for this chapter, do not code to use design pattern; when you code and see a good fit, only then use design pattern.

Now, we will go onto testing, which is a must for any serious development effort.

6
Test-Driven Development

In this chapter, we will discuss some good concepts that are to be applied during testing. First, we will take a look at how we can create mock or stubs easily to test functionalities that are not present in the system. Then, we will cover how to write test cases with parameterization. Custom test runners can be of great help to write test utilities for a specific project. Then, we will cover how to test threaded applications, and utilize concurrent execution to decrease the overall time spent on test suite runs. We will cover the following topics:

- Mock for tests
- Parameterization
- Creating custom test runners
- Testing threaded applications
- Running test cases in parallel

Mock for tests

Key 1: Mock what you do not have.

When we are using test driven development, we have to write test cases for the components that rely on other components that are not written yet or take a lot of time to execute. This is close to impossible until we create mocks and stubs. In this scenario, stubs or mocks are very useful. We use a fake object instead of a real one to write the test case. This can be made very easy if we use tools that are provided by the language. For example, in the following code, we only have the interface for the worker class, and no real implementation of it. We want to test the `assign_if_free` function.

Instead of writing any stub ourselves, we use the `create_autospec` function to create a mock object from the definition of the Worker abstract class. We also set up a return value for the function call of checking whether worker was busy or not:

```python
import six
import unittest
import sys
import abc
if sys.version_info[0:2] >= (3, 3):
    from unittest.mock import Mock,
create_autospec else:
    from mock import Mock, create_autospec
if six.PY2:
    import thread
else:
    import _thread as thread


class IWorker(six.with_metaclass(abc.ABCMeta, object)):

    @abc.abstractmethod
    def execute(self, *args):
        """ execute an api task """
        pass

    @abc.abstractmethod
    def is_busy(self):
        pass

    @abc.abstractmethod
    def serve_api(self,):
        """register for api hit"""
        pass


class Worker(IWorker):
    def __init__(self,):
        self.__running = False

    def execute(self,*args):
        self.__running = True
        th = thread.start_new_thread(lambda x:time.sleep(5))
        th.join()
        self.__running = False

    def is_busy(self):
```

```
            return self.__running == True

    def assign_if_free(worker, task):
        if not worker.is_busy():
            worker.execute(task)
            return True
        else:
            return False


class TestWorkerReporting(unittest.TestCase):

    def test_worker_busy(self,):
        mworker = create_autospec(IWorker)
        mworker.configure_mock(**{'is_busy.return_value':True})
        self.assertFalse(assign_if_free(mworker, {}))

    def test_worker_free(self,):
        mworker = create_autospec(IWorker)
        mworker.configure_mock(**{'is_busy.return_value':False})
        self.assertTrue(assign_if_free(mworker, {}))


if __name__ == '__main__':
    unittest.main()
```

To set up return values, we can also use functions to return conditional responses, as follows:

```
>>> STATE = False
>>> worker = create_autospec(Worker,)
>>> worker.configure_mock(**{'is_busy.side_effect':lambda : True if
not STATE else False})
>>> worker.is_busy()
True
>>> STATE=True
>>> worker.is_busy(
) False
```

We can also set methods to raise exceptions using the side_effect attribute of mock, as follows:

```
>>> worker.configure_mock(**{'execute.side_effect':Exception('timeout
for execution')})
```

>>>

```
>>> worker.execute()
Traceback (most recent call last):
  File "<stdin>", line 1, in <module>
  File "/usr/lib/python3.4/unittest/mock.py", line 896, in
    __call__ return _mock_self._mock_call(*args, **kwargs)
  File "/usr/lib/python3.4/unittest/mock.py", line 952, in
    _mock_call raise effect
Exception: timeout for execution
```

Another use is to check whether a method was called and with what arguments, as follows:

```
>>> worker = create_autospec(IWorker,)
>>> worker.configure_mock(**{'is_busy.return_value':True})
>>> assign_if_free(worker,{})
False
>>> worker.execute.calle
d False
>>> worker.configure_mock(**{'is_busy.return_value':False})
>>> assign_if_free(worker,{})
True
>>> worker.execute.called
True
```

Parameterization

Key 2: Manageable inputs to tests.

For the tests where we have to test various inputs for the same functionality or transformations, we have to write test cases to cover test different inputs. We can use parameterization here. In this way, we invoke the same test case with different inputs, hence, decreasing time and errors that are associated with it. Newer Python versions 3.4 or higher include a very useful method, subTest in unittest. TestCase, which makes it very easy to add parameterized tests. In the test output, please note that the parameterized values are also available:

```
import unittest
from itertools import combinations
from functools import wraps

def convert(alpha):
```

```
    return ','.join([str(ord(i)-96) for i in alpha])

class TestOne(unittest.TestCase):

    def test_system(self,):
        cases = [("aa","1,1"),("bc","2,3"),("jk","4,5"),("
xy","24,26")]
        for case in cases:
            with self.subTest(case=case):
                self.assertEqual(convert(case[0]),case[1])

if __name__ == '__main__':
    unittest.main(verbosity=2)
```

This will give us the following output:

```
(py3)arun@olappy:~/codes/projects/pybook/book/ch6$ python
parametrized.py
test_system (__main__.TestOne) ...
======================================================================
FAIL: test_system (__main__.TestOne) (case=('jk', '4,5'))
----------------------------------------------------------------------
Traceback (most recent call last):
  File "parametrized.py", line 14, in test_system
    self.assertEqual(convert(case[0]),case[1])
AssertionError: '10,11' != '4,5'
- 10,11
+ 4,5


======================================================================
= FAIL: test_system (__main__.TestOne) (case=('xy', '24,26')) -------
----------------------------------------------------------------------
Traceback (most recent call last):
  File "parametrized.py", line 14, in test_system
    self.assertEqual(convert(case[0]),case[1])
AssertionError: '24,25' != '24,26'
- 24,25
?    ^  +
24,26
?^


----------------------------------------------------------------------
Ran 1 test in 0.001s

FAILED (failures=2)
```

This also means that if we needed the *currying* that is running tests for all combinations of inputs, then this can be done very easily. We have to write a function that returns curried arguments, and then we can use subTest to have mini tests run with curried arguments. This way it is very easy to explain to new people on the team how to write test cases with minimum language jargon, as follows:

```python
import unittest
from itertools import combinations
from functools import wraps

def entry(number,alpha):
    if 0 < number < 4 and 'a' <= alpha <= 'c':
        return True
    else:
        return False

def curry(*args):
    if not args:
        return []
    else:
        cases = [ [i,] for i in args[0]]
        if len(args)>1:
            for i in range(1,len(args)):
                ncases = []
                for j in args[i]:
                    for case in cases:
                        ncases.append(case+[j,])
                cases = ncases
        return cases


class TestOne(unittest.TestCase):

    def test_sample2(self,):
        case1 = [1,2]
        case2 = ['a','b','d']
        for case in curry(case1,case2):
            with self.subTest(case=case):
                self.assertTrue(entry(*case), "not equal")

if __name__ == '__main__':
    unittest.main(verbosity=2)
```

This will give us the following output:

```
(py3)arun@olappy:~/codes/projects/pybook/book/ch6$ python
parametrized_curry.py
test_sample2 (__main__.TestOne) ...
==================================================================
= FAIL: test_sample2 (__main__.TestOne) (case=[1, 'd']) -------------
---------------------------------------------------- Traceback
(most recent call last):
  File "parametrized_curry.py", line 33, in test_sample2
    self.assertTrue(entry(*case), "not equal")
AssertionError: False is not true : not equal


==================================================================
= FAIL: test_sample2 (__main__.TestOne) (case=[2, 'd']) -------------
---------------------------------------------------- Traceback
(most recent call last):
  File "parametrized_curry.py", line 33, in test_sample2
    self.assertTrue(entry(*case), "not equal")
AssertionError: False is not true : not equal


---------------------------------------------------------------------
- Ran 1 test in 0.000s

FAILED (failures=2)
```

But, this works only for new versions of Python. For the older versions, we can perform similar work using dynamism of language. We can implement this feature ourselves, as shown in the following code snippet. We use a decorator to stick the parameterize value to test case, and then in `metaclass`, we create a new wrapper function that calls the original function with the required parameters:

```
from functools import wraps
import six
import unittest
from datetime import datetime, timedelta

class parameterize(object):
    """decorator to pass parameters to function we
    need this to attach parameterize arguments on
    to the function, and it attaches
    __parameterize_this__ attribute which tells
    metaclass that we have to work on this
    attribute """
```

```
    def __init__(self,names,cases):
        """ save parameters """
        self.names = names
        self.cases = cases

    def __call__(self,func):
        """ attach parameters to same func """
        func.__parameterize_this__ = (self.names, self.cases)
        return func

class ParameterizeMeta(type):

    def __new__(metaname, classname, baseclasses, attrs):
        # iterate over attribute and find out which one have
__ parameterize_this__ set
        for attrname, attrobject in six.iteritems(attrs.copy()):
            if attrname.startswith('test_'):
                pmo = getattr(attrobject,'__parameterize_this__',None)
                if pmo:
                    params,values = pmo
                    for case in values:
                        name = attrname + '_'+'_'.join([str(item) for
item in case])
                        def func(selfobj,
testcase=attrobject,casepass =dict(zip(params,case))):
                            return testcase(selfobj, **casepass)
                        attrs[name] = func
                        func.__name__ = name
                    del attrs[attrname]
        return type.__new__(metaname, classname, baseclasses, attrs)

class MyProjectTestCase(six.with_metaclass(ParameterizeMeta,unittest.
TestCase)):
    pass

class TestCase(MyProjectTestCase):

    @parameterize(names=("input","output"),
                  cases=[(1,2),(2,4),(3,6)])
    def test_sample(self,input,output):
        self.assertEqual(input*2,output)


    @parameterize(names=("in1","in2","output","shouldpass"),
```

```
                    cases=[(1,2,3,True),
                           (2,3,6,False)]
                  )
    def test_sample2(self,in1,in2,output,shouldpass):
        res = in1 + in2 == output
        self.assertEqual(res,shouldpass)


if __name__ == '__main__':
    unittest.main(verbosity=2)
```

The output for the preceding code is as follows:

```
test_sample2_1_2_3_True (__main__.TestCase) ... ok
test_sample2_2_3_6_False (__main__.TestCase) ...
ok test_sample_1_2 (__main__.TestCase) ... ok
test_sample_2_4 (__main__.TestCase) ... ok
test_sample_3_6 (__main__.TestCase) ... ok


----------------------------------------------------------------------

- Ran 5 tests in 0.000s

OK
```

Creating custom test runners

Key 3: Getting information from test system.

The flow of unit test is like this: unittest TestProgram in unittest.main is the primary object that runs everything. Test cases are collected by test discovery or by loading modules that were passed via command line. If no test runner is specified to the main function, by default, TextTestRunner is used. Test suite is passed to the runner's run function to give back a TestResult object.

The custom test runners are a great way to get information in a specific output format, from the test system, manage run sequence, store results in a database, or create new features for project needs.

Let's now take a look at an example to create an XML output of test cases, you may need something like this to integrate with continuous integration systems, which are only able to work with some XML format. As in the following code snippet `XMLTestResult` is the class that gives the test result in the XML format. The `TsRunner` class test runner then puts the same information on the `stdout` stream. We also add the time taken for the test case as well. The `XMLify` class is sending information to test the `TsRunner` runner class in an XML format. The `XMLRunner` class is putting this information in the XML format on `stdout`, as follows:

```
""" custom test system classes """

import unittest
import sys
import time
from xml.etree import ElementTree as ET
from unittest import TextTestRunner


class XMLTestResult(unittest.TestResult):
    """converts test results to xml format"""

    def __init__(self, *args,**kwargs):#runner):
        unittest.TestResult.__init__(self,*args,**kwargs
        ) self.xmldoc = ET.fromstring('<testsuite />')

    def startTest(self, test):
        """called before each test case run"""
        test.starttime = time.time()
        test.testxml = ET.SubElement(self.xmldoc,
                                     'testcase',
                                     attrib={'name': test._
testMethodName,
                                            'classname': test.__
class__.__name__,
                                            'module': test.__
module__})

    def stopTest(self, test):
        """called after each test case"""
        et = time.time()
        time_elapsed = et - test.starttime
        test.testxml.attrib['time'] = str(time_elapsed)

    def addSuccess(self, test):
```

```
        """
        called on successful test case run
        """
        test.testxml.attrib['result'] = 'ok'

    def addError(self, test, err):
        """
        called on errors in test case
        :param test: test case
        :param err: error info
        """
        unittest.TestResult.addError(self, test, err)
        test.testxml.attrib['result'] = 'error'
        el = ET.SubElement(test.testxml, 'error', )
        el.text = self._exc_info_to_string(err, test)

    def addFailure(self, test, err):
        """
        called on failures in test cases.
        :param test: test case
        :param err: error info
        """
        unittest.TestResult.addFailure(self, test, err)
        test.testxml.attrib['result'] = 'failure'
        el = ET.SubElement(test.testxml, 'failure', )
        el.text = self._exc_info_to_string(err, test)

    def addSkip(self, test, reason):
        # self.skipped.append(test)
        test.testxml.attrib['result'] = 'skipped'
        el = ET.SubElement(test.testxml, 'skipped',
        ) el.attrib['message'] = reason


class XMLRunner(object):
    """ custom runner class"""

    def __init__(self, *args,**kwargs):
        self.resultclass = XMLTestResult

    def run(self, test):
        """ run given test case or suite"""
        result = self.resultclass()
        st = time.time()
```

```
        test(result)
        time_taken = float(time.time() - st)
        result.xmldoc.attrib['time'] = str(time_taken)

        ET.dump(result.xmldoc)
        #tree = ET.ElementTree(result.xmldoc)
        #tree.write("testm.xml", encoding='utf-8')
        return result
```

Let's assume that we use this XMLRunner on the test cases, as shown in the following code:

```python
import unittest

class TestAll(unittest.TestCase):
    def test_ok(self):
        assert 1 == 1

    def test_notok(self):
        assert 1 >= 3

    @unittest.skip("not needed")
    def test_skipped(self):
        assert 2 == 4

class TestAll2(unittest.TestCase):
    def test_ok2(self):
        raise IndexError
        assert 1 == 1

    def test_notok2(self):
        assert 1 == 3

    @unittest.skip("not needed")
    def test_skipped2(self):
        assert 2 == 4

if __name__ == '__main__':
    from ts2 import XMLRunner
unittest.main(verbosity=2, testRunner=XMLRunner)
```

We will get the following output:

```
<testsuite    time="0.0005891323089599609"><testcase    classname="TestAll"
module="__main__"  name="test_notok"  result="failure"  time="0.000237703
3233642578"><failure>Traceback (most recent call last):
  File "test_cases.py", line 8, in test_notok
    assert 1 &gt;= 3
AssertionError
</failure></testcase><testcase classname="TestAll" module="__
main__" name="test_ok" result="ok" time="2.6464462280273438e-05"
/><testcase classname="TestAll" module="__main__" name="test_skipped"
result="skipped" time="9.059906005859375e-06"><skipped message="not
needed" /></testcase><testcase classname="TestAll2" module="__
main__" name="test_notok2" result="failure" time="9.34600830078125e-
05"><failure>Traceback (most recent call last):
  File "test_cases.py", line 20, in
    test_notok2 assert 1 == 3
AssertionError
</failure></testcase><testcase classname="TestAll2" module="__
main__" name="test_ok2" result="error" time="8.440017700195312e-
05"><error>Traceback (most recent call last):
  File "test_cases.py", line 16, in test_ok2
    raise IndexError
IndexError
</error></testcase><testcase classname="TestAll2" module="__main__"
name="test_skipped2" result="skipped" time="7.867813110351562e-
06"><skipped message="not needed" /></testcase></testsuite>
```

Testing threaded applications

Key 4: Make threaded application tests like nonthreaded ones.

My experience with testing on threaded application is to perform the following actions:

- Try to make the threaded application as nonthreaded as possible for tests. By this, I mean that group logic that is nonthreaded in one code segment. Do not try to test business logic with thread logic. Try to keep them separate.
- Work with as little global state as possible. Functions should pass around objects that are needed to work.
- Try to make queues of tasks to synchronize them. Instead of creating producer consumer chains yourself, first try to use queues.
- Also note that sleep statements make test cases run slower. If you add up sleeps in the code for more than 20 places, the whole test suite starts to become slow. Threaded code should pass on information with events and notifications rather than a while loop checking some condition.

The `_thread` module in Python 2 and the `_thread` module in Python 3 are a big help as you can start functions as threads, shown as follows:

```
>>> def foo(waittime):
...     time.sleep(waittime)
...     print("done")
>>> thread.start_new_thread(foo,(3,))
140360468600576
>> done
```

Running test cases in parallel

Key 5: Faster test suite execution

When we have accumulated a lot of test cases in the project, it takes a lot of time to execute all of the test cases. We have to make the test run in parallel to decrease the time that is taken overall. In this case, the `py.test` testing framework does a fantastic job of simplifying the ability to run tests in parallel. To make this work, we have to first install the `py.test` library, and then use its runner to run the test cases. The `py.test` library has an `xdist` plugin, which adds the capability to run tests in parallel, as follows:

```
(py35) [ ch6 ] $ py.test -n 3 test_system.py
======================================= test session starts =======
========================================
platform linux -- Python 3.5.0, pytest-2.8.2, py-1.4.30, pluggy-0.3.1
rootdir: /home/arun/codes/workspace/pybook/ch6, inifile: plugins:
xdist-1.13.1
gw0 [5] / gw1 [5] / gw2 [5]
scheduling tests via LoadScheduling
s...F
============================================== FAILURES
============ ===================================
_____ TestApi.test_api2

[gw0] linux -- Python 3.5.0 /home/arun/.pyenv/versions/py35/bin/
python3.5
self = <test_system.TestApi testMethod=test_api2>

    def test_api2(self,):
        """api2
            simple test1"""
        for i in range(7):
            with self.subTest(i=i):
```

```
>                   self.assertLess(i, 4, "not less")
E                   AssertionError: 4 not less than 4 : not less

test_system.py:40: AssertionError
=============================== 1 failed, 3 passed, 1 skipped in 0.42
seconds ================================
```

If you want to dive deeper into this topic, you can refer to `https://pypi.python.org/pypi/pytest-xdist`.

Summary

Testing is very important in creating a stable application. In this chapter, we discussed how we mock the objects to create an easy separation on concerns to test different components. Parameterization is very useful to test various transformation logics. The most important take away is to try to create functionalities that are needed by your project as test utilities. Try to stick with the `unittest` module. Use other libraries for parallel execution as they support the `unittest` tests as well.

In the next chapter, we will cover optimization techniques for Python.

7
Optimization Techniques

In this chapter, we will learn how to optimize our Python code to get better responsive programs. But, before we dive into this, I would like to stress that do not optimize until it is necessary. A better-readable program has a better life and maintainability than a tersely-optimized program. First, we will take a look at simple optimization tricks to keep a program optimized. We should have knowledge about them so that we can apply easy optimizations from the start. Then, we will look at profiling to find bottlenecks in the current program and apply optimizations where we need them. As a last resort, we can compile in the C language and provide functionality as an extension to Python. Here is the gist of topics that we will cover:

- Writing optimized code
- Profiling to find bottlenecks
- Using fast libraries
- Using C speeds

Writing optimized code
Key 1: Easy optimizations for code.

We should pay close attention to not use loops inside loops, giving us quadratic behavior. We can use built-ins, such as map, ZIP, and reduce, instead of using loops if possible. For example, in the following code, the one with map is faster because the looping is implicit and done at C level. By plotting their run times respectively on graph as test 1 and test 2, we see that it is nearly constant for PyPy but reduces a lot for CPython, as follows:

```
def sqrt_1(ll):
    """simple for loop"""
    res = []
```

```
    for i in ll:
        res.append(math.sqrt(i))
    return res

def sqrt_2(ll):
    "builtin map"
    return list(map(math.sqrt,ll))
The test 1 is for sqrt_1(list(range(1000))) and test
22 sqrt_2(list(range(1000))).
```

The following image is a graphical representation of the preceding code:

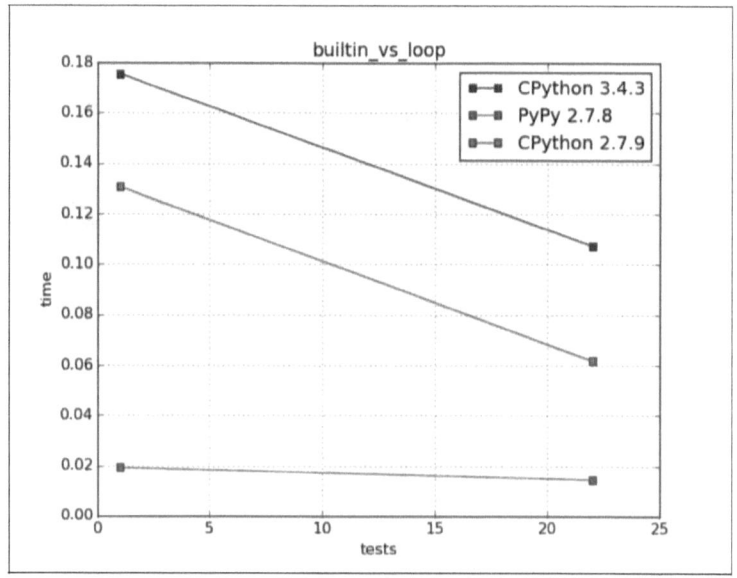

Generators should be used, when the result that is consumed is averagely smaller than the total result consumed. In other words, the result that is generated in the end may not be used. They also serve to conserve memory because no temporary result is stored but generated on demand. In the following example, sqrt_5 creates a generator, while sqrt_6 creates a list. The use_combo instance breaks out of the loop of iteration after a given number of iterations. Test 1 runs use_combo(sqrt_5,range(10),5) and all results are consumed from iterator, whereas test 2 is for the use_combo(sqrt_6,range(10),5) generator. Test 1 should take more time than test 2 as it creates results for all ranges of inputs. Tests 3, and 4 are run with a range of 25, and tests 5, and 6 are run with a range of 100. As it can be seen, the time consumption variation increases with no of elements in the list:

```
def sqrt_5(ll):
    "simple for loop, yield"
    for i in ll:
        yield i,math.sqrt(i)

def sqrt_6(ll):
    "simple for loop"
    res = []
    for i in ll:
        res.append((i,math.sqrt(i)))
    return res

def use_combo(combofunc,ll,no):
    for i,j in combofunc(ll):
        if i>no:
            return j
```

The following image is the graphical representation of the preceding code:

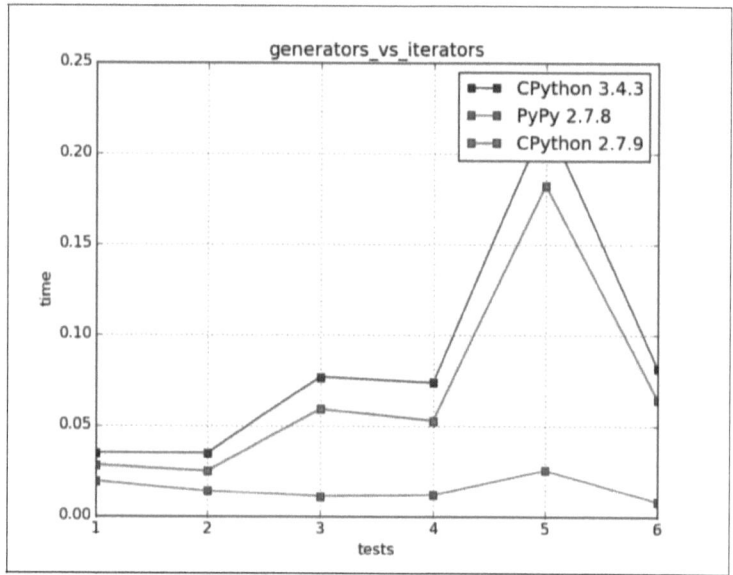

When we are inside a loop and reference an outside namespace variable, it is first searched in local, then nonlocal, followed by global, and then built-in scopes. If the number of repetitions are more, then such overheads add up. We can reduce namespace lookup by making such global/built-in objects available in the local namespace. For example, in the following code snippet, `sqrt_7(test2)` will be faster than `sqrt_1(test1)` because of the same reasons:

```
def sqrt_1(ll):
    """simple for loop"""
    res = []
    for i in ll:
        res.append(math.sqrt(i))
    return res

def sqrt_7(ll):
    "simple for loop, local"
    sqrt = math.sqrt
```

```
res = []
for i in ll:
    res.append(sqrt(i))
return res
```

The following image is the graphical representation of the preceding code:

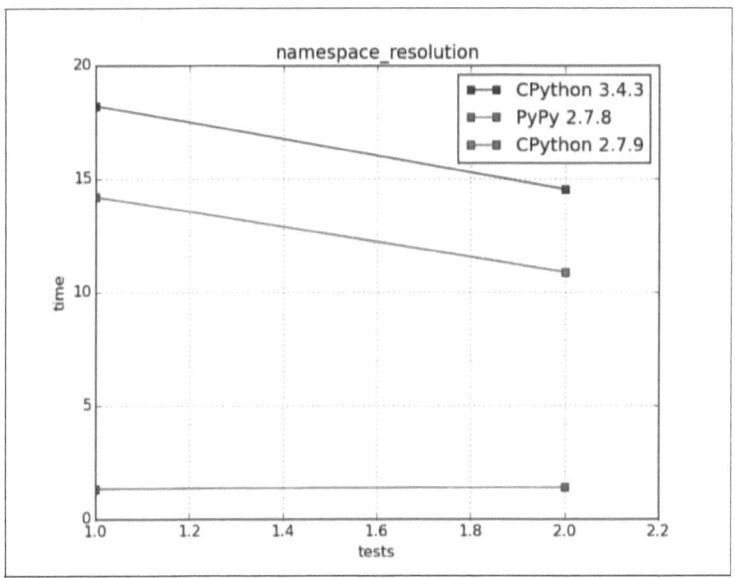

The cost of subclassing is not much and subclassing doesn't make method calls slower even if common sense says that it will take lot of time to look a method up on the inheritance hierarchy. Let's take the following example:

```
class Super1(object):
    def get_sqrt(self,no):
        return math.sqrt(no)

class Super2(Super1):
```

```
        pass

    class Super3(Super2):
        pass

    class Super4(Super3):
        pass

    class Super5(Super4):
        pass

    class Super6(Super5):
        pass

    class Super7(Super6):
        pass

    class Actual(Super7):
        """method resolution via hierarchy"""
        pass

    class Actual2(object):
        """method resolution single step"""
        def get_sqrt(self,no):
            return math.sqrt(no)

    def use_sqrt_class(aclass,ll):
        cls_instance = aclass()
        res = []
        for i in ll:
            res.append(cls_instance.get_sqrt(i))
        return res
```

Here, if we call `get_sqrt` on the `Actual(case1)` class, we need to search it seven levels deep in its base classes, whereas for the `Actual2(case2)` class it is present on the class itself. The following graph is our plot for both scenarios:

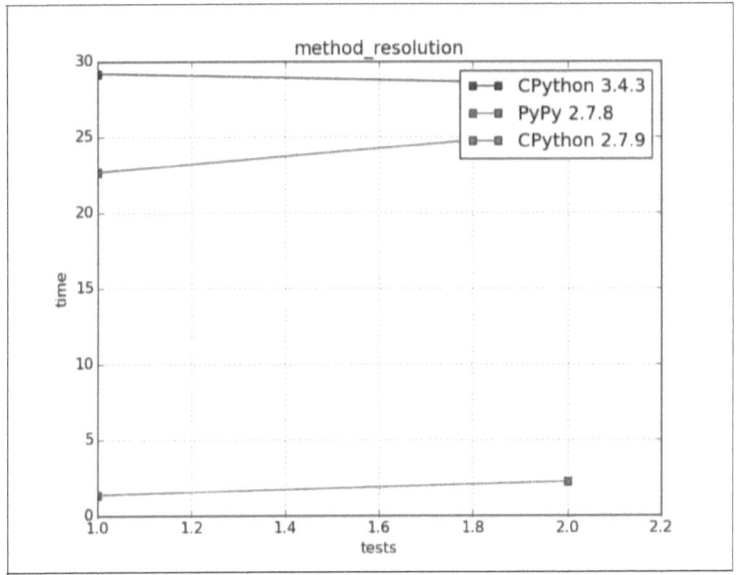

Also, if we are using too many checks in the program logic for return codes or error conditions, we should see how many such checks are really needed. We can write the program logic without using any checks and then get the errors in the exception handling logic. This makes the code easy to understand. As in the following example, the `getf_1` function uses checks to filter out error conditions, but too many checks are making code hard to understand. The other `get_f2` function is the same application logic or algorithm with exception handling. For test 1 (`get_f1`) and test 2 (`get_f2`), no file is present, so all exceptions are raised. In this scenario, the exception handling logic, that is test 2, takes more time. For test 3 (`get_f1`) and test 4 (`get_f2`), the file and key are present; hence, no error is raised. In this case, test 4 takes less time, as follows:

```
def getf_1(ll):
    "simple for loop,checks"
```

```
    res = []
    for fname in ll:
        curr = []
        if os.path.exists(fname):
            f = open(fname,"r")
            try:
                fdict = json.load(f)
            except (TypeError, ValueError):
                curr = [fname,None,"Unable to read
            Json"] finally:
                f.close()
            if 'name' in fdict:
                curr = [fname,fdict["name"],'']
            else:
                curr = [fname,None,"Key not found in file"]
        else:
            curr = [fname,None,"file not found"]
        res.append(curr)
    return res

def getf_2(ll):
    "simple for loop, try-except"
    res = []
    for fname in ll:
        try:
            f = open(fname,"r")
            res.append([fname,json.load(f)['name'],''])
        except IOError:
            res.append([fname,None,"File Not Found Error"])
        except TypeError:
            res.append([fname,None,'Unable to read Json'])
        except KeyError:
            res.append([fname,None,'Key not found in file'])
        except Exception as e:
            res.append([fname,None,str(e)])
        finally:
            if 'f' in locals():
                f.close()
    return res
```

The following image is the graphical representation of the preceding code:

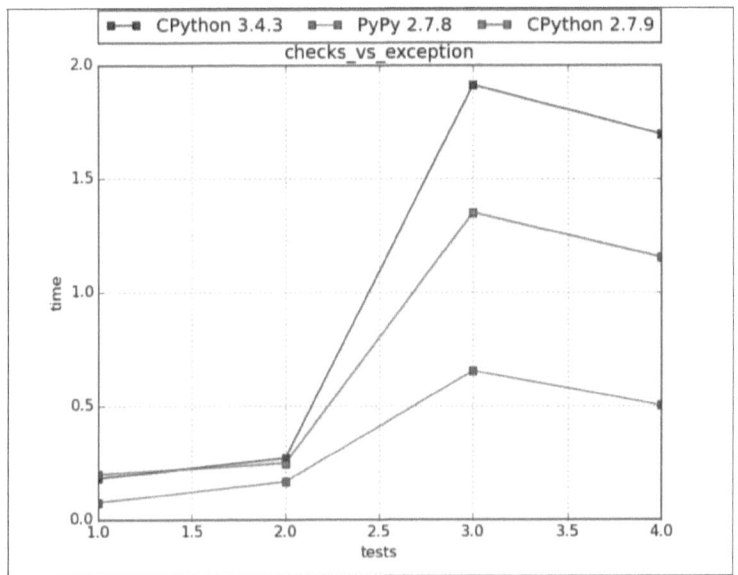

Function calling has overheads and if the performance bottlenecks can be removed by reducing function calls, we should do so. Typically, functions call in loops. In the following example, when we wrote logic inline, it took less time. Also, for PyPy such effects are less in general as most called functions in loops are generally called with the same type of arguments; hence, they get compiled. Any further call to these functions is like calling a C language function:

```
def please_sqrt(no):
    return math.sqrt(no)

def get_sqrt(no):
    return please_sqrt(no)

def use_sqrt1(ll,no):
    for i in ll:
        res = get_sqrt(i)
        if res >= no:
```

```
            return i

def use_sqrt2(ll,no):
    for i in ll:
        res = math.sqrt(i)
        if res >= no:
            return i
```

The following image is the graphical representation of the preceding code:

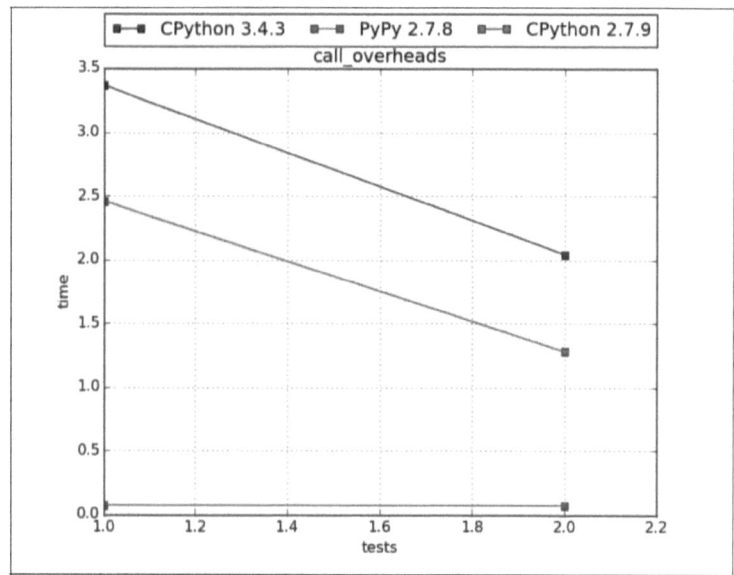

Profiling to find bottlenecks

Key 2: Identifying application performance bottlenecks.

We should not rely on our intuition on how to optimize application. There are two major ways for logic slowdown; one is CPU time taken, and the second is the wait for results from some other entity. By profiling, we can find out such cases in which we can tweak logic, and language syntax to get better performance on the same hardware. The following code is a `showtime` decorator that I use to calculate the time taken to call a function. It is simple and effective to get rapid answers:

```python
from datetime import datetime,timedelta
from functools import wraps
import time

def showtime(func):

    @wraps(func)
    def wrap(*args,**kwargs):
        st = time.time() #time clock can be used too
        result = func(*args,**kwargs) et =
        time.time()
        print("%s:%s"%(func.__name__,et-st))
        return result
    return wrap

@showtime
def func_c():
    for i in range(1000):
        for j in range(1000):
            for k in range(100):
                pass

if __name__ == '__main__':
    func_c()
```

This will give us the following output:

```
(py35) [ ch7 ] $ python code_1_8.py
func_c:1.3181400299072266
```

When profiling a single large function that does a lot of stuff, we may need to know on what particular line we are spending the most time. This query can be answered using the `line_profiler` module. You can get it with `pip install line_profiler`. It shows the time that is spent per line. To get results, we should decorate the function with a special profile decorator that will be used by `line_profiler`:

```
from datetime import datetime,timedelta
from functools import wraps
import time
import line_profiler

l = []
def func_a():
    global l
    for i in range(10000):
        l.append(i)

def func_b():
    m = list(range(100000))

def func_c():
    func_a()
    func_b()
    k = list(range(100000))

if __name__ == '__main__':
    profiler = line_profiler.LineProfiler()
    profiler.add_function(func_c)
    profiler.run('func_c()')
    profiler.print_stats()
```

This will give us the following output:

```
Timer unit: 1e-06 s

Total time: 0.007759 s
File: code_1_9.py
Function: func_c at line 15

Line #      Hits         Time  Per Hit   % Time  Line Contents
==============================================================
    15                                           def func_c():
    16         1         2976   2976.0     38.4      func_a()
    17         1         2824   2824.0     36.4      func_b()
    18         1         1959   1959.0     25.2      k =
list(range(100000))
```

Another way of profiling is using the `kernprof` program that is supplied with the `line_profiler` module. We have to decorate the function to be a profiler by the `@profile` decorator and run the program, as shown in the following code snippet:

```
from datetime import datetime,timedelta
from functools import wraps
import time

l = []
def func_a():
    global l
    for i in range(10000):
        l.append(i)

def func_b():
    m = list(range(100000))

@profile
def func_c():
    func_a()
    func_b()
    k = list(range(100000))
```

The output for this will be as follows:

```
(py35) [ ch7 ] $ kernprof -l -v code_1_10.py
Wrote profile results to code_1_10.py.lprof
Timer unit: 1e-06 s

Total time: 0 s
File: code_1_10.py
Function: func_c at line 14

Line #      Hits         Time  Per Hit   % Time  Line Contents
==============================================================
    14                                           @profile
    15                                           def func_c():
    16                                               func_a()
    17                                               func_b()
    18                                               k =
list(range(100000))
```

Memory profilers are a very good tool to estimate memory consumption in a program. To profile a function, simply decorate it with profile and run the program like this:

```
from memory_profiler import profile

@profile(precision=4)
def sample():
    l1 = [ i for i in range(10000)]
    l2 = [ i for i in range(1000)]
    l3 = [ i for i in range(100000)]
    return 0

if __name__ == '__main__':
    sample()
```

To get details on the command line, use the following code:

```
(py36)[ ch7 ] $ python  ex7_1.py
Filename: ex7_1.py

Line #      Mem usage    Increment  Line Contents
================================================
     8     12.6 MiB      0.0 MiB    @profile
     9                              def sample():
    10     13.1 MiB      0.5 MiB        l1 = [ i for i in range(10000)]
    11     13.1 MiB      0.0 MiB        l2 = [ i for i in range(1000)]
    12     17.0 MiB      3.9 MiB        l3 = [ i for i in
range(100000)]
    13     17.0 MiB      0.0 MiB        return 0


Filename: ex7_1.py

Line #      Mem usage    Increment  Line Contents
================================================
    10     13.1 MiB      0.0 MiB        l1 = [ i for i in range(10000)]


Filename: ex7_1.py

Line #      Mem usage    Increment  Line Contents
```

```
====================================================
    12      17.0 MiB      0.0 MiB        13 = [ i for i in
range(100000)]
```

```
Filename: ex7_1.py

Line #    Mem usage    Increment   Line Contents
====================================================
    11      13.1 MiB      0.0 MiB        12 = [ i for i in range(1000)]
```

We can also use it to debug long-running programs. The following code is for a simple socket server. It adds lists to the global lists variable, which never gets deleted. Saving contents in `simple_serv.py` is as follows:

```python
from SocketServer import BaseRequestHandler,TCPServer

lists = []

class Handler(BaseRequestHandler):
    def handle(self):
        data = self.request.recv(1024).strip()
        lists.append(list(range(100000)))
        self.request.sendall("server got "+data)

if __name__ == '__main__':
    HOST,PORT = "localhost",9999
    server = TCPServer((HOST,PORT),Handler)
    server.serve_forever()
```

Now, run the program via profiler as follows:

```
mprof run simple_serv.py
```

Put some bogus hits to the server. I used the `netcat` utility:

```
[ ch7 ] $ nc localhost 9999 <<END
hello
END
```

Kill the server after some time and plot the memory consumed over time with the following code:

```
[ ch7 ] $ mprof plot
```

We get a good graph showing us memory consumption over time:

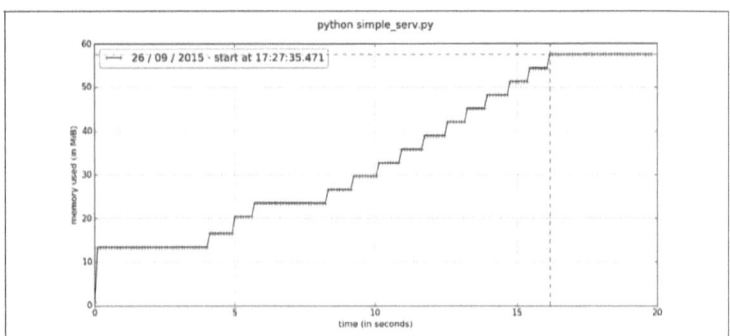

Other than getting program memory consumption, we may be interested in objects carrying spaces. Objgraph (`https://pypi.python.org/pypi/objgraph`) is able to graph object links for your programs. Guppy (`https://pypi.python.org/pypi/guppy/`) is another package that has heapy, which is a heap analysis tool. It is very helpful to see the number of objects on heap for a running program. As of this writing, it was only available for Python 2. For analysis of a long-running process, Dowser (`https://pypi.python.org/pypi/dowser`) is also a good choice. We can use Dowser to see the memory consumption to run Celery or a WSGI server. Django-Dowser is good and provides the same functionality as an app, but as the name suggests, it only works with Django.

Using fast libraries

Key 3: Use easy drop-in faster libraries.

There are libraries out there that can help a lot in optimizing code, rather than writing some optimized routines yourself. For example, if we have a list that needs to be fast at FIFO, we may use the `blist` package. We can use C versions of libraries, such as `cStringIO` (faster StringIO), `ujson` (faster JSON handling), numpy (math, and vectors), and `lxml` (XML handling). Most of the libraries that are listed here are just a Python wrapper over C libraries. You only need to search once for your problem domain. Other than this, we can make a C, or C++ library interface with Python very easily, which is also our next topic.

Using C speeds

Key 4: Running at C speeds.

SWIG

SWIG is an interface compiler that connects programs written in C, and C++ with scripting languages. We can use SWIG to call C, C++ compiled in Python. Let's say that we have a factorial computing library in C, with source code in the `fact.c` file and the corresponding `fact.h` header file:

The source code in `fact.c` is as follows:

```
#include "fact.h"
long int fact(long int n) {
    if (n < 0){
        return 0;
    }
    if (n == 0) {
        return 1;
    }
    else {
        return n * fact(n-1);
    }
}
```

The source code in `fact.h` is as follows:

```
long int fact(long int n);
```

Now, we need to write an interface file for SWIG, which tells it what it needs to be exposed to Python:

```
/* File: fact.i */
%module fact
%{
#define SWIG_FILE_WITH_INIT
#include "fact.h"
%}
long int fact(long int n);
```

Here, module indicates the module name for the Python library, and SWIG_FILE_WITH_INIT indicates that the resulting C code should be built with a Python extension. The content in {% %} is used in the C wrap code that is generated. We have three files, fact.c, fact.h, and fact.i, in directory. We run SWIG to generate wrapper_code as follows:

```
swig3.0 -python -O -py3 fact.i
```

The -O option is used for optimizations and -py3 is for Python 3 specific features.

This generates fact.py and fact_wrap.c. The fact.py is a Python module and fact_wrap.c is the glue code between C and Python:

```
gcc -fpic -c fact_wrap.c fact.c -
I/home/arun/.virtualenvs/py3/include/ python3.4m
```

Here, I have to include my python.h path to compile it. This will generate fact.o and fact_wrap.o. Now, the last part is to create a dynamic linked library, as follows:

```
gcc -shared fact.o fact_wrap.o -o _fact.so
```

The _fact.so file is used by the fact.py to run C functions. Now, we can use the fact module in our Python programs:

```
>>> from fact import fact
>>> fact(10)
3628800
>>> fact(5)
120
>>> fact(20)
2432902008176640000
```

CFFI

The **C Foreign Function Interface (CFFI)** for Python is one tool that looks the best to me because of the easy setup and interface. It works on an ABI and API level.

Using our factorial C programs here as well, we first create a shared library for the code:

```
gcc -shared fact.o -o _fact.so
gcc -fpic -c fact.c -o fact.o
```

Now, we have a `_fact.so` shared library object in our current directory. To load this in the Python environment, we can perform this action which is very straightforward. We should have header files for the library so that we can use declarations. Install the CFFI package from distribution or pip that is needed for this, as follows:

```
>>> from cffi import FFI
>>> ffi = FFI()
>>> ffi.cdef("""
... long int fact(long int num);
... """)
>>> C = ffi.dlopen("./_fact.so")
>>> C.fact(20)
2432902008176640000
```

We can reduce import times for the module if we do not call `cdef` in the import modules. We can write another `setup_fact_ffi.py` module that gives us a `fact_ffi.py` module with the compiled information. Hence, the load times decrease a lot:

```
from cffi import FFI

ffi = FFI()
ffi.set_source("fact_ffi", None)
ffi.cdef("""
    long int fact(long int n);
""")

if __name__ == "__main__":
ffi.compile()

python setup_fact_ffi.py
```

We now can use this module to get `ffi` and load our shared library as follows:

```
>>> from fact_ffi import ffi
>>> ll = ffi.dlopen("./_fact.so")
>>> ll.fact(20)
2432902008176640000
```
>>>

Until this point, as we were using a precompiled shared library, we didn't need a compiler. Let's suppose that there is this small C function that you need in Python, and you do not want to write another .c file for it, then this is how it can be done. You can also extend it to shared libraries as well.

First, we define a `build_ffi.py` file, which will compile and create a module for us:

```
__author__ = 'arun'

from cffi import FFI
ffi = FFI()

ffi.set_source("_fact_cffi",
    """
    long int fact(long int n) {
    if (n < 0){
        return 0;
    }
    if (n == 0) {
        return 1;
    }
    else {
        return n * fact(n-1);
    }
}
""",
                libraries=[]
    )

ffi.cdef("""
long int fact(long int n);
""")

if __name__ == '__main__':
    ffi.compile()
```

When we run Python `fact_build.py`, this will create a `_fact_cffi.cpython-34m.so` module. To use it, we have to import it and use the `lib` variable to access the module:

```
>>> from _fact_cffi import ffi,lib
>>> lib.fact(20)
```

Cython

Cython is like a superset of Python in which we can optionally give static declarations. The source code is compiled to C/C++ extension modules.

We write our old factorial program in `fact_cpy.pyx` as follows:

```
cpdef double fact(int num):
    cdef double res = 1
    if num < 0:
        return -1
    elif num == 0:
        return 1
    else:
        for i in range(1,num + 1):
            res = res*i
    return res
```

Here `cpdef` is the function declaration for CPython that creates a Python function and conversion logic for arguments, and a C function that actually executes. `cpdef` is defining the data type for the res variable, which helps in speedup.

We have to create a `setup.py` file to compile this code into an extension module (we can directly use it by using `pyximport` but we will leave that for now). The contents for the `setup.py` file will be as follows:

```
from distutils.core import setup
from Cython.Build import cythonize

setup(
  name = 'factorial library',
    ext_modules = cythonize("fact_cpy.pyx"),
    )
```

Now to build the module, all we have to do is type in the following command, and we get a `fact_cpy.cpython-34m.so` file in the current directory:

```
python setup.py build_ext --inplace
```

Using this in Python is as follows:

```
>>> from fact_cpy import fact
>>> fact(20)
2432902008176640000
```

Summary

In this chapter, we saw various techniques that are used to optimize and profile code. I will, again, point out that we should always focus on first writing the correct program, then writing test cases for it, and then optimizing it. We should write code with optimizations that we know at that time or without optimization the first time, and we should hunt for them only if we need them from a business perspective. Compiling a C module can give a good speedup for CPU-intensive tasks. Also, we can give up GIL in C modules, which can also help us in increasing performance. But, all of this was on single system. Now, in the next chapter, we will see how we can improve performance when the tricks that were discussed in this chapter are not sufficient for a real-life scenario.

8
Scaling Python

In this chapter, we will try to understand how we can make our program work for more inputs by making the program scalable. We will do this by both optimizing and adding computing power to the system. We will cover the following topics:

- Going multithreaded
- Using multiple processes
- Going asynchronous
- Scaling horizontally

The major reason a system is not able to scale is state. Events can change the state of a system permanently for both that request or further requests from that endpoint.

Normally state is stored in the database, and reactions to events are worked on sequentially, and changes to state due to events are then stored in DB.

Task can be computation intensive (CPU load) or IO bound in which system needs answers from some other entity. Here, `taskg` and `taskng` are GIL and nonGIL versions of a task. The `taskng` task is in C module compiled via SWIG by enabling its threads:

```
URL = "http://localhost:8080/%s"
def cputask(num,gil=True):
    if gil:
        return taskg(num)
    else:
        return taskng(num)
def iotask(num):
    req = urllib.request.urlopen(URL%str(num))
    text = req.read()
    return text
```

For example, I have created a test server that responds to requests after 1 sec. To create a comparison for scenarios, we first create a simple serial program. As expected, both IO and CPU tasks time add up:

```
import time
from tasker import cputask, iotask
from random import randint
def process(rep, case=None):
        inputs = [[randint(1, 1000), None] for i in range(rep)
    ] st = time.time()

    if 'cpu' == case:
        for i in inputs:
            i[1] = cputask(i[0])
    elif 'io' == case:
        for i in inputs:
            i[1] = iotask(i[0])
    tot = time.time() - st
    for i in inputs:
        assert i[0] == int(i[1]), "not same %s" % (i)
    return tot
```

Outputs can be easily summarized as follows:

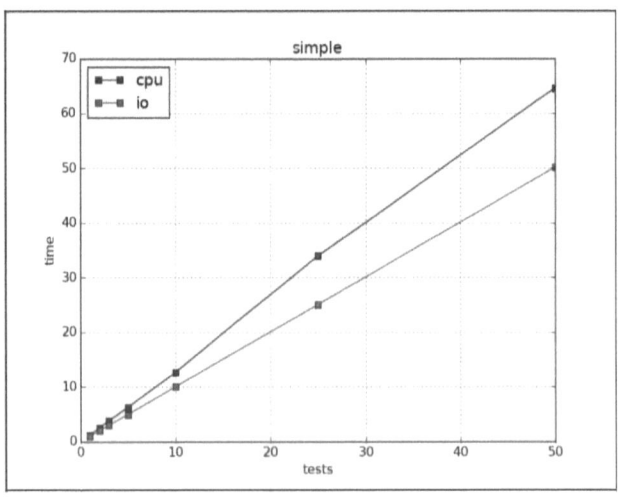

Going multithreaded

Key 1: Using threads to process in parallel.

Let's see how threads can help us in improving performance. In Python, due to Global Interpreter Lock, only one thread runs at a given time. Also, context is switched as all of them are given a chance to run. Hence, this is load in addition to computation. Hence, CPU-intensive tasks should take the same or more time. IO tasks are not doing anything but waiting, so they will get the boost. In the following code segment, `threaded_iotask` and `threaded_cputask` are two functions that are executed using separate threads. The code is run for various values to get results. The process function invokes multiple threads for tasks and sums up the timings taken:

```python
import time
from tasker import cputask, iotask
from random import randint
import threading,random,string

def threaded_iotask(i):
    i[1] = iotask(i[0])

def threaded_cputask(i):
    i[1] = cputask(i[0])

stats = {}

def process(rep, cases=()):
    stats.clear()
    inputs = [[randint(1, 1000), None] for i in range(rep)
    ] threads = []
    if 'cpu' in cases:
        threads.extend([
            threading.Thread(target=threaded_cputask, args=(i,))
                for i in inputs])
    elif 'io' in cases:
        threads.extend([
            threading.Thread(target=threaded_iotask, args=(i,))
                for i in inputs])
    stats['st'] = stats.get('st',time.time())
    for t in threads:
        t.start()
    for t in threads:
        t.join()
    stats['et'] = stats.get('et',time.time())
```

```
tot = stats['et']  - stats['st']
for i in inputs:
    assert i[0] == int(i[1])
return tot
```

Plotting various results onscreen, we can easily see that threading is helping IO tasks but not CPU tasks:

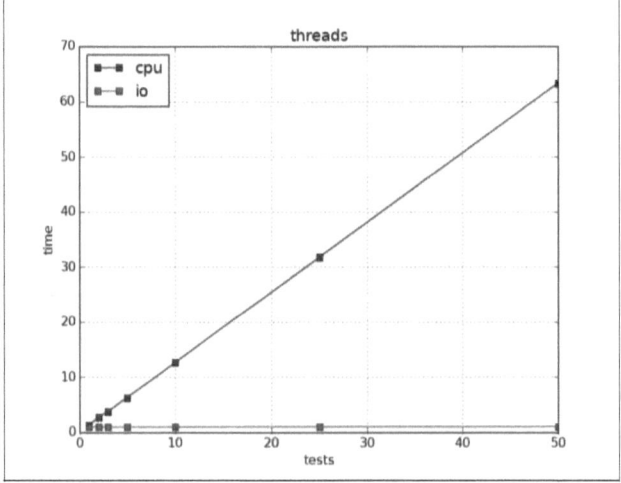

As discussed, this is due to GIL. Our CPU task is defined in C, we can give up GIL to see whether that helps. The following is plot of run with no GIL for tasks. We can see that CPU tasks are now taking a lot less time than before. But, GIL is there for a reason. If we give up GIL, atomicity for data structures is not guaranteed as there may be two threads working on same data structure at a given time.

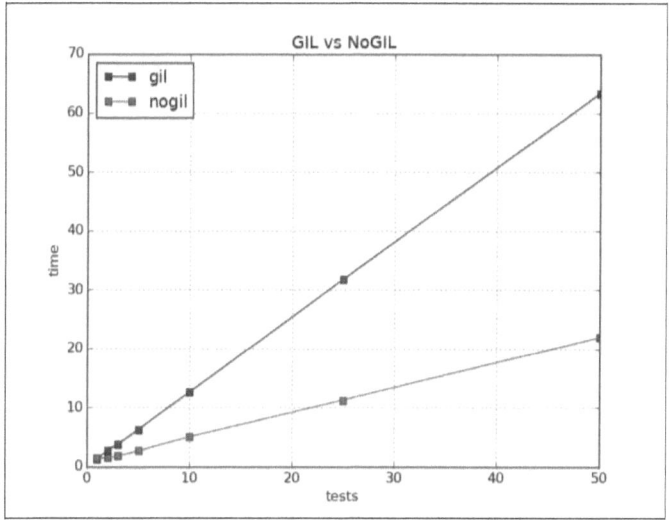

Using multiple processes

Key 2: Churning CPU-intensive tasks.

Multiple processes are helpful to fully utilize all CPU cores. It helps in CPU-intensive work as tasks are run in separate processes, and there is no GIL between actual working processes. The setup and communication cost between processes is higher than threads. In the following code section, proc_iotask, proc_cputask are processes that run them for various inputs:

```
import time
from tasker import cputask, iotask
from random import randint
```

```
import multiprocessing,random,string

def proc_iotask(i,outq):
    i[1] = iotask(i[0])
    outq.put(i)

def proc_cputask(i,outq):
    res = cputask(i[0])
    outq.put((i[0],res))

stats = {}

def process(rep, case=None):
    stats.clear()
    inputs = [[randint(1, 1000), None] for i in range(rep)
    ] outq = multiprocessing.Queue() processes = []
    if 'cpu' == case:
        processes.extend([
            multiprocessing.Process(target=proc_cputask,
args=(i,outq))
                for i in inputs])
    elif 'io' == case:
        processes.extend([
            multiprocessing.Process(target=proc_iotask, args=(i,outq))
                for i in inputs])
    stats['st'] = stats.get('st',time.time())
    for t in processes:
        t.start()
    for t in processes:
        t.join()
    stats['et'] = stats.get('et',time.time())
    tot = stats['et'] - stats['st']
    while not outq.empty():
        item = outq.get()
        assert item[0] == int(item[1])
    return tot
```

In the following diagram, we can see the multiple IO operations are getting a boost from multiprocessing. CPU tasks are also getting a boost from multiprocessing:

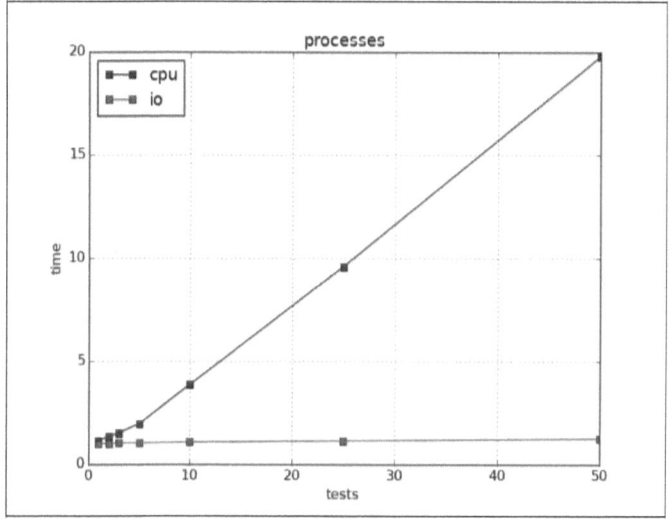

If we compare all four: serial, threads, threads without GIL, and multiprocesses, we will observe that threads without GIL and multiprocesses are taking almost the same time. Also, serial and threads are taking the same time, which shows little benefit of using threads in CPU-intensive tasks:

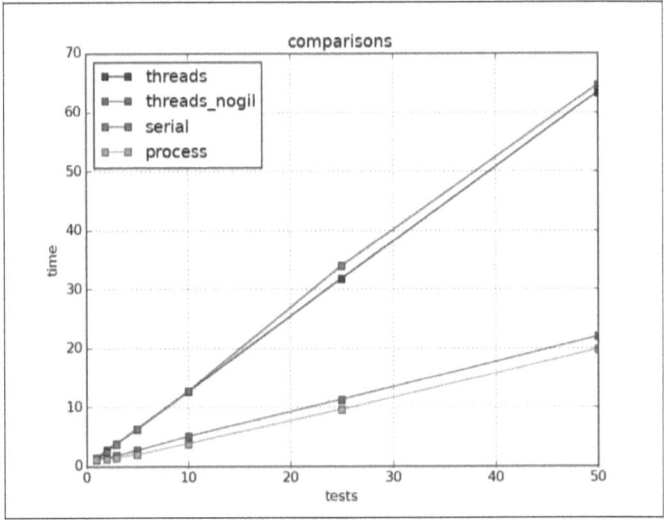

Going asynchronous

Key 3: Being asynchronous for parallel execution.

We can also process more than one request by being asynchronous as well. In this method, instead of us polling for updates from objects, they tell us when they have a result. Hence, the main thread in the meantime can execute other stuff. Asyncio, Twisted, and Tornado are libraries in Python that can help us write such code.

Asyncio, and Tornado are supported in Python 3, and some portions of Twisted also run on Python 3 as of now. Python 3.5 introduced the `async` and `await` keywords that helps write asynchronous code. The `async` keyword defines that the function is an asynchronous function and that the result may not be available right away. The `await` keyword waits until the results are captured and returns the result.

In the following code, `await` in the main function waits for all the results to be available:

```
import time, asyncio
from tasker import cputask, async_iotask
from random import randint
import aiopg, string, random, aiohttp
from asyncio import futures, ensure_future,
gather from functools import partial

URL = "http://localhost:8080/%s"

async def async_iotask(num, loop=None):
    res = await aiohttp.get(URL % str(num[0]), loop=loop)
    text = await res.text()
    num[1] = int(text)
    return text

stats = {}

async def main(rep, case=None, loop=None, inputs=None):
    stats.clear()
    stats['st'] = time.time()
    if 'cpu' == case:
        for i in inputs:
            i[1] = cputask(i[0])
    if 'io' == case:
        deferreds = []
        for i in inputs:
            deferreds.append(async_iotask(i, loop=loop))
        await gather(*deferreds, return_exceptions=True, loop=loop)
    stats['et'] = time.time()


def process(rep, case=None):
    loop = asyncio.new_event_loop()
    inputs = [[randint(1, 1000), None] for i in range(rep) ]
    loop.run_until_complete(main(rep, case=case, loop=loop,
inputs=inputs))
    loop.close()
    tot = stats['et'] - stats['st']
    # print(inputs)
    for i in inputs:
        assert i[0] ==
    int(i[1]) return tot
```

Plotting results on the graph, we can see that we got a boost on the IO portion, but for CPU-intensive work, it takes time similar to serial:

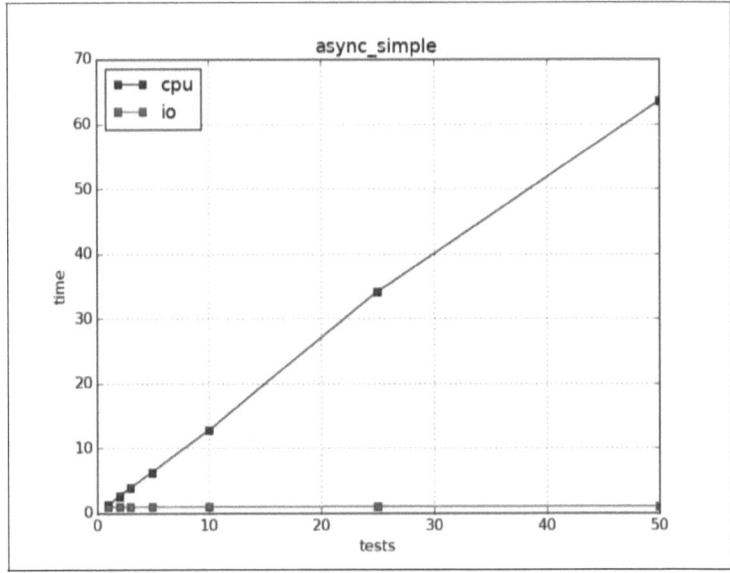

CPU tasks are blocking everything, hence, this is a bad design. We have to use either threads, or better multiprocessing to help in CPU-intensive tasks. To run tasks in threads or processes, we can use `ThreadPoolExecutor,` and `ProcessPoolExecutor` from the `concurrent.futures` package. The following is the code for `ThreadPoolExecutor:`

```
async def main(rep,case=None,loop=None,inputs=[]):
    if case == 'cpu':
        tp = ThreadPoolExecutor()
        futures = []
        for i in inputs:
            task = partial(threaded_cputask,i)
            future = loop.run_in_executor(tp,task)
            futures.append(future)
        res = await asyncio.gather(*futures,return_
exceptions=True,loop=loop)
```

For `ProcessPoolExecutor,` we have to use a multiprocessing queue to collect results back, as follows:

```
def threaded_cputask(i,outq):
    res = cputask(i[0])
    outq.put((i[0],res))

async def main(rep,case=None,loop=None,outq=None,inputs=[]):
    if case == 'cpu':
        pp = ProcessPoolExecutor()
        futures = []
        for i in inputs:
            task = partial(threaded_cputask,i,outq)
            future = loop.run_in_executor(pp,task)
            futures.append(future)
        res = await asyncio.gather(*futures,return_
exceptions=True,loop=loop)

def process(rep,case=None):
    loop = asyncio.new_event_loop()
    inputs = [[randint(1, 1000), None] for i in range(rep)
    ] st = time.time()
    m = multiprocessing.Manager()
    outq = m.Queue()
    loop.run_until_complete(main(rep,case=case,loop=loop,outq=outq,in
puts=inputs))
    tot = time.time() - st
    while not outq.empty():
        item = outq.get()
        assert item[0] == int(item[1])
    loop.close()
    return tot
```

Plotting the results, we can see that threads are taking more or less the same time as without them, but they still help make the program more responsive as the program will be able to perform other IO tasks in the meantime. Multiprocessing gives the max boost:

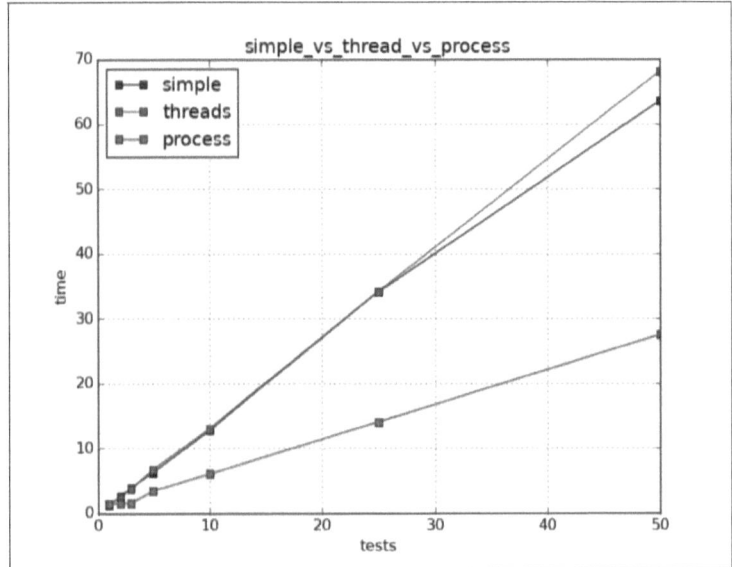

Async systems are used mostly when IO is the main thing. As you can see, it is similar to serial for CPU. Let's now take a look at which one is better, threading or `async`, for our scalable IO-based application. We used the same IO task but on higher loads. Asyncio gives failures and takes more time than threads. I tested this on Python 3.5:

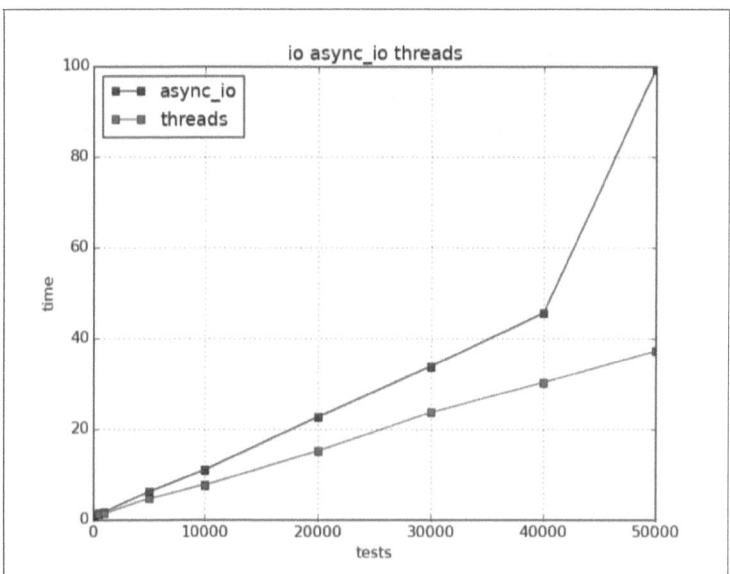

The last advice will be to look at other implementations as well, such as PyPy, Jython, IronPython, and so on.

Scaling horizontally

If we add further nodes to the application, it must add to the total processing power. To create frontend systems that perform more data transmission than computation, `async` frameworks are better suited. If we use PyPy, it will give a performance boost to the application. Code for Python 3 or Python 2 compatibility using six or other such libraries so that we can use anything available for optimization.

We can use message pack or JSON for message transfer. I prefer JSON for language agnostic and easy-text representation. Workers can be multiprocessing workers for CPU-bound tasks or thread-based for other scenarios.

The system should not store the state but pass it with messages. Everything doesn't need to be in DB. We can take some things out when not necessary.

ZeroMQ (messageQueue): ZMQ is a wonderful library that acts as a glue to connect your programs together. It has connectors for almost all language. You can easily use multiple languages/frameworks to enable their communication with ZMQ and among themselves. It also provides tools to create various utilities. Let's now look at how we can create a load-balanced worker system easily using ZMQ. In the following code snippet, we created a client (requester) that can ask for a result from a group of servers (workers) that are load balanced. In the following code, we can see the socket type is DEALER. Sockets in ZMQ can be thought of as mini servers. The req sockets do not actually transmit until they get a response for the previous one. DEALER and ROUTER sockets are better suited for real-life scenarios. The code for synchronization is as follows:

```python
import sys
import zmq
from zmq.eventloop import ioloop
from zmq.eventloop.ioloop import IOLoop
from zmq.eventloop.zmqstream import
ZMQStream ioloop.install()


class Cli():

    def __init__(self, name, addresses):
        self.addresses = addresses
        self.loop = IOLoop.current()
        self.ctx = zmq.Context.instance()
        self.skt = None
        self.stream = None
        self.name = bytes(name, encoding='ascii')
        self.req_no = 0
        self.run()

    def run(self):
        self.skt = self.ctx.socket(zmq.DEALER)
        for address in self.addresses:
            self.skt.connect(address)
```

```
            self.stream = ZMQStream(self.skt)
            self.stream.on_recv(self.handle_request)
            self.loop.call_later(1, self.send_request)

        def send_request(self):
            msg = [self.req_no.to_bytes(1, 'little'), b"hello"]
            print("sending", msg)
            self.stream.send_multipart(msg)
            self.req_no += 1
            if self.req_no < 10:
                self.loop.call_later(1, self.send_request)

        def handle_request(self, msg):
            print("received", int.from_bytes(msg[0], 'little'), msg[1])

    if __name__ == '__main__':
        print("starting  client")
        loop = IOLoop.current()
        serv = Cli(sys.argv[1], sys.argv[2:])
        loop.start()
```

The following is the code for servers or actual workers. We can have many of them and the load is distributed in a round-robin fashion among them:

```
import sys

import zmq
from zmq.eventloop import ioloop
from zmq.eventloop.ioloop import IOLoop
from zmq.eventloop.zmqstream import

ZMQStream ioloop.install()


class Serv():

    def __init__(self, name, address):
        self.address = address
        self.loop = IOLoop.current()
        self.ctx = zmq.Context.instance()
        self.skt = None
        self.stream = None
        self.name = bytes(name, encoding='ascii')
```

```
            self.run()

    def run(self):
        self.skt = self.ctx.socket(zmq.ROUTER)
        self.skt.bind(self.address)
        self.stream = ZMQStream(self.skt)

        self.stream.on_recv(self.handle_request)

    def handle_request(self, msg):
        print("received", msg)
        self.stream.send_multipart(msg)


if __name__ == '__main__':
    print("starting server")
    serv = Serv(sys.argv[1], sys.argv[2])
    loop = IOLoop.current()
    loop.start()
```

The following is the result from the run:

```
For client
(py35) [ scale_zmq ] $ python client.py "cli" "tcp://127.0.0.1:8004"
"tcp://127.0.0.1:8005"
starting  client
sending [b'\x00', b'hello']
sending [b'\x01', b'hello']
received 1 b'hello'
sending [b'\x02', b'hello']
sending [b'\x03', b'hello']
received 3 b'hello'
sending [b'\x04', b'hello']
sending [b'\x05', b'hello']
received 5 b'hello'
sending [b'\x06', b'hello']
sending [b'\x07', b'hello']
received 7 b'hello'
sending [b'\x08', b'hello']
sending [b'\t', b'hello']
received 9 b'hello'
received 0 b'hello'
received 2 b'hello'
received 4 b'hello'
```

```
received 6 b'hello'
received 8 b'hello'

Outputs server/workers:
(py35) [ scale_zmq ] $ python server.py "serv" "tcp://127.0.0.1:8004"
starting server
received [b'\x00k\x8bEg', b'\x00', b'hello']
received [b'\x00k\x8bEg', b'\x02', b'hello']
received [b'\x00k\x8bEg', b'\x04', b'hello']
received [b'\x00k\x8bEg', b'\x06', b'hello']
received [b'\x00k\x8bEg', b'\x08', b'hello']
(py35) [ scale_zmq ] $ python server.py "serv" "tcp://127.0.0.1:8005"
starting server
received [b'\x00k\x8bEg', b'\x01', b'hello']
received [b'\x00k\x8bEg', b'\x03', b'hello']
received [b'\x00k\x8bEg', b'\x05', b'hello']
received [b'\x00k\x8bEg', b'\x07', b'hello']
received [b'\x00k\x8bEg', b'\t', b'hello']
```

We can use the third-party package Supervisord to make workers restart on failure.

The real power of ZMQ is in creating network architecture and nodes as required by the project from simpler components. You can test the framework easily as it can support IPC, TCP, UDP, and many more protocols. They can also be used interchangeably.

There are other libraries/frameworks as well that can help a lot in this space, such as NSQ, Python parallel. Many projects go for RabbitMQ as the broker and AMQP as the protocol. Choosing good communication is very important for the design and scalability of a system, and it depends on the project requirement.

Summary

Making a program scalable is easy if we separate portions of program and use each part tuned for best performance. In this chapter, we saw how various portions of Python help in vertical as well as horizontal scaling. All this information must be taken into consideration when designing architecture of the application.

Index

A

abstract classes
  using 35
abstract factory 84-86
adaptor pattern 77, 78
algorithms
  about 63
  implementing 63-66
application performance bottlenecks
  identifying 117-122
arrays 59
asynchronous operations
  performing, for parallel execution 136-141
attributes
  about 14
  class method 17, 18
  descriptors 16, 17
  instance method 17, 18
  searching 14-16
  static method 17, 18

B

binary tree 59, 60 built-in
data structures 51-55

C

callable objects
  replacing 40-44
C Foreign Function Interface (CFFI) 124-126
class
  language protocols, using 32

class inheritance
  about 29
  Method Resolution Order (MRO) 29
  superclass's methods, calling 30
class method 17, 18
class objects
  creating 12, 13
code optimization 107-116
collections.deque 57
command pattern 82-84
comprehensions 45
context manager protocol 34, 35
custom test runners
  creating 99-103
Cython 126, 127

D

decorators 40
descriptors 16, 17
design patterns
  abstract factory 84-86
  adaptor pattern 77, 78
  command pattern 82-84
  facade pattern 79, 80
  flyweight pattern 80-82
  observer pattern 70, 71
  registry pattern 86-88
  singleton pattern 74, 75
  state pattern 88, 89
  strategy pattern 72-74
  template pattern 75, 76
dictionary 56
Dowser
  URL 122

static method 17, 18
strategy pattern 72-74
superclass
  method, calling 30
SWIG 123, 124

T

template pattern 75, 76
test cases
  running, in parallel 104, 105
tests
  mocks, creating for 91-94
third party data structures
  about 59
  arrays 59
  binary tree 59, 60
  list 59
  sorted containers 60, 61
  trie 61, 62

threaded applications
  testing 103, 104
tuples 55

U

utilities
  about 45, 46
  itertools utility 46

X

xdist plugin
  about 104
  URL 105

Z

ZeroMQ (messageQueue) 142

NumPy: Beginner's Guide

Third Edition

ISBN: 978-1-78528-196-9   Paperback: 348 pages

Build efficient, high-speed programs using the high-performance NumPy mathematical library

1. Written as a step-by-step guide, this book aims to give you a strong foundation in NumPy and breaks down its complex library features into simple tasks.

2. Perform high performance calculations with clean and efficient NumPy code.

3. Analyze large datasets with statistical functions and execute complex linear algebra and mathematical computations.

NumPy Cookbook

Second Edition

ISBN: 978-1-78439-094-5   Paperback: 258 pages

Over 90 fascinating recipes to learn and perform mathematical, scientific, and engineering Python computations with NumPy

1. Perform high-performance calculations with clean and efficient NumPy code.

2. Simplify large data sets by analyzing them with statistical functions.

3. A solution-based guide packed with engaging recipes to execute complex linear algebra and mathematical computations.

Please check **www.PacktPub.com** for information on our titles

www.ingramcontent.com/pod-product-compliance
Lightning Source LLC
Chambersburg PA
CBHW032018170526
45157CB00002B/754